MW00531997

Venture Investing in Science

Venture Investing in Science

Douglas W. Jamison
and Stephen R. Waite

foreword by **Mark Anderson**

Columbia Business School

Columbia University Press
Publishers Since 1893
New York   Chichester, West Sussex
cup.columbia.edu
Copyright © 2017 Columbia University Press
All rights reserved

Library of Congress Cataloging-in-Publication Data
Names: Jamison, Douglas W., author. | Waite, Stephen R., author.
Title: Venture investing in science / Douglas W. Jamison and Stephen R. Waite.
Description: New York : Columbia University Press, [2017] | Series: Columbia
Business School Publishing | Includes bibliographical references and index.
Identifiers: LCCN 2016047015 | ISBN 9780231175722 (cloth : alk. paper) |
ISBN 9780231544702 (e-book)
Subjects: LCSH: Venture capital—United States. | New business
enterprises—United States—Finance. | Science and industry—United States.
Classification: LCC HG4751 b .J364 2017 | DDC 332.6—dc23
LC record available at https://lccn.loc.gov/2016047015

Columbia University Press books are printed on permanent
and durable acid-free paper.
Printed in the United States of America

Cover design: Fifth Letter

To Charles E. Harris, mentor and kindred spirit.

Contents

Foreword by Mark Anderson   ix

Preface   xiii

Acknowledgments   xvii

Introduction   1

1
Deep Science Disruption   18

2
The U.S. Deep Science Innovation Ecosystem   50

3
Deep Science and the Evolution
of American Venture Capital   80

Contents

4
Diversity Breakdown in Venture Investing   109

5
Fostering Diversity in Venture Investing   132

6
Deep Science Venture Investing   168

7
Our Choice Ahead   185

Appendix 1:
The Case of D-Wave Systems   205

Appendix 2:
The Case of Nantero   233

Notes   249

Index   263

Foreword

MARK ANDERSON

There are probably as many venture investing models today as there are venture capitalists around the world, yet we can rather easily categorize their interests and effects by their targets. Unfortunately, we are living in an era when the conventional icons, target industries, and risk allocations common in the past have changed to include short-term, often spurious (if lucrative) plays.

The hallmarks of this transition are apparent in the growing forest of unicorn ($1 billion or more) start-up valuations for companies that are based on little more than social touch, with no revenues beyond inflated advertising statistics. An early leading indicator of the real risk inherent to this new category of behavior was the entry, and then driving influence, of a group of venture firms that made their deals without any apparent regard to price.

While no good capitalist criticizes another's attempt at maximizing return on investment, an obvious schism is arising between short-term investments into profitless global social apps, for example, and what Douglas Jamison and Stephen Waite term "deep science investing." What is the financial, economic, social, or even political difference between putting a dollar into Twitter versus the first cancer vaccine? What is the difference in value or return between investing in an app that originated as a way of ranking coeds at Harvard (the app that led to Facebook) versus Leroy Hood's work building the machines that decode DNA sequences? Certainly more short-term money has come from the mix of privacy invasion and advertising income at Facebook, yet experts wonder whether such apps actually contribute to a slide in productivity. Just as surely, more long-term returns will come from the thousands of firms now doing genetic research and linking this knowledge to medicine.

A series of major shifts occurring in the world are leading to strong divisions of investor self-interest, large-scale splits between both economic activities and outcomes, and the types and terms of benefits that flow from these activities. We can easily identify the difference between science and technology that enhances production and that which enhances consumption. In the computer world, for example, people who make things use personal computers, workstations, and laptops; people who consume things use tablets and smartphones.

Is it possible that the majority of investors are doing nothing more than utilizing a vast network of glass, servers, and apps for the sole sake of selling advertising to a burgeoning consumer culture?

This same kind of schism exists in the kind of science we are doing today, compared with that of a century or so ago. With many great exceptions, science today seems incremental, focused more on peer review and acquiring team grants as opposed to making waves and revolutionary discoveries. Charts such as those in this book paint a picture of movement and focus over the past hundred years, from discoveries on the order of quantum mechanics to applications of science in the world of technology and product development. Are we moving from science to tech, from the fundamental to the incremental, from tectonic shifts in our basic knowledge to dusting off the silverware?

It is impossible to imagine being alive in 1905, during what is called Einstein's *annus mirabilis*. During that single year, this relatively young and unknown patent examiner published four papers that changed science forever, launching the fields of quantum mechanics, special relativity, nuclear science, and the power of statistical science in proving the existence of atoms.

Does that sound like any year the present-day reader remembers? Not likely. If we note that the same decade or two also included the contributions of scientists like Neils Bohr, Louis de Broglie, Max Planck, Werner Heisenberg, and Paul Dirac—well, one wonders where are their intellectual descendants. Without diving into the deeper questions of how so many brilliant scientists and their discoveries could have clustered at that moment, or why such clusterings are less common, there is no question regarding the importance of working to enhance our movement toward this caliber of deep science, nor for its value to both human knowledge and society.

Fortunately there are great scientists and entrepreneurs at work today. Indeed, having left an era when pure academics (Planck, Heisenberg) or rank amateurs (Newton, Galileo) were driving scientific discovery, the schisms previously noted have led to the modern scientist/entrepreneur, a special class of brilliant individuals who have the courage to ignore their peers' lack of understanding or support and who are willing to start their own companies as a path to getting the word out.

This is the opportunity landscape now facing venture investors, whether they be individual angels, venture firms, private equity funds, or global corporations seeking advantage. On a daily basis, venture firms choose between investing in "hard science" areas (such as quantum computing, materials science, nuclear or alternative energy, precision medicine, non–von Neumann architectures, and deep learning) and "soft" areas (like social networking, gaming, or even TV content). In *Venture Investing in Science*, Jamison and Waite have crystalized the rationale behind the massive benefits of venture investing in deep science. It isn't easy, it isn't usually quick, and it isn't for everyone, as they point out early on. But investors can be sorted in just the same way as they sort their own targets: by their ultimate goals.

For investors with the intellect and patience to attain highest and best values for their money, long-term investing in deep science is the path to follow.

Preface

This book is the product of several decades of experience in venture investing in what we refer to in this book as "deep science." During this time, we have found, built, invested in, worked with, and taken public deep science companies in the U.S. public markets. Additionally, we have done so from an investment vehicle that is itself a U.S. microcapitalization public company. We have had a front-row seat observing the trends highlighted and described in the book. One need not be a venture capitalist or have any practical experience as a venture investor to read and appreciate this book. We believe the subject matter of this book is deserving of a wide audience, including entrepreneurs, business executives, and policymakers. It is written in a manner to serve such an audience.

Our main thesis is straightforward. Over the past decade, there has been a breakdown in the vital process of commercializing deep science–based inventions and innovation that fuel economic dynamism in the United States. Venture capital, long a critical ingredient of the deep science commercialization process in the United

States since the end of World War II, has increasingly migrated away from deep science toward software investments. This shift has produced a diversity breakdown in venture capital investing.

The diversity breakdown in venture capital is occurring during a period of fundamental changes in the structure and workings of U.S. public capital markets. In the aggregate, these capital market shifts have inhibited the funding of early-stage deep science–based companies. The result has been a waning of economic dynamism essential to fostering higher living standards over time.

This book consists of seven chapters and two appendices featuring real-world case studies of venture investing in deep science. The introduction provides a general overview of the book's main thesis. Chapters 1, 2, and 3 provide a historical context for readers related to advances in deep science and technological change and their profound impact on economic development, as well as the genesis and evolution of venture capital in the United States. The historical context, which may be skimmed by those intimately familiar with the history of science, technological change, and venture capital, is provided to facilitate a better understanding and appreciation of the main thesis.

Chapter 4 discusses the diversity breakdown in venture capital, and chapter 5 illuminates the shifts in U.S. public capital markets that have worked to impede venture investing in deep science and that have dampened American economic dynamism. In chapter 6, we examine some of the emerging trends in deep science that could provide growth and prosperity if we find a way to bring diversity back to venture capital investing.

In the final chapter, we summarize the book's main thesis and discuss some potential implications for the future. The appendices include two case studies in deep science venture investing: D-Wave Systems and Nantero.

It is our hope that this book provides a foundation for further discussion on the importance of venture investing in deep science to U.S. economic dynamism and also to global economic dynamism.

While we discuss some ominous investing and economic trends, our natures push us toward a belief in rational optimism, hope, and seeing a reversal of fortunes that will foster prosperity and the general welfare of American society. We encourage those sympathetic with the main thesis of this book who have an interest in helping to spur venture investing in deep science to contact us. We can be reached by email: VIISBook@gmail.com.

Acknowledgments

The authors wish to thank our publisher, Myles Thompson, for the opportunity to collaborate on this book with Columbia University Press. Our editor, Stephen Wesley, was invaluable in helping to prepare the manuscript for publication. We are grateful for his assistance. Our colleagues at the Harris & Harris Group have been supportive of the work that has gone into researching and writing this book. We thank each of them. We also wish to thank D-Wave Systems CEO Vern Brownell and Nantero cofounder and CEO Greg Schmergel for their cooperation and assistance with the case studies published in the appendices. Our families have been a constant source of support and encouragement. There are no words to express thanks for all that you are and do.

Venture Investing in Science

Introduction

Venture Investing in Deep Science

> But without scientific progress no amount of achievement in other directions can insure our health, prosperity, and security as a nation in the modern world.
> —VANNEVAR BUSH

THE 1950S WAS A period of rebellion and experimentation in the United States that emanated from the confluence of a post–World War II euphoria and the paranoia that the Cold War brought. Out of this period came the rise of powerful technology based on science and backed by venture capital. The emergence of venture capital–backed technology rooted in science changed the course of innovation and business, while fundamentally altering the structure of the economy over time. Among the powerful new technologies to emerge were the computer and other electronic devices that ultimately changed how people lived, worked, and played.

It is only fitting that the first great venture capital success arose during this time: Digital Equipment Corporation (DEC), founded in 1957. DEC was born from funding provided by the Department of Defense to the Lincoln Laboratory at the Massachusetts Institute of Technology (MIT). The lab was established to build new computer technology for the Cold War. DEC blossomed because of the adoption of another great deep science advance in the 1950s, the transistor, developed by three engineers at Bell Labs.

DEC's founders, Kenneth Olsen and Harlan Anderson, believed the use of the transistor could lead to faster, more efficient, and hence smaller computers. The minicomputer was born, built to compete against the large, expensive, vacuum tube–based mainframe computers pioneered by International Business Machines (IBM).

The first American venture capital firm, American Research and Development (ARD), invested $70,000 into DEC in 1957 and took control of a 70 percent stake in the company. In 1966, DEC completed its initial public offering (IPO), raising $8 million, making ARD's initial $70,000 investment five hundred times more valuable at the IPO price. By the summer of 1967, DEC stock had gone from $22 at its IPO to over $110. The first venture capital–backed entrepreneurial company "home run" was born. The minicomputer was the beginning of what would become a revolution in computing. This revolution continues today with the introduction of new mobile and augmented-reality computing technologies. Deep science was translated into technology, into innovation, and into great wealth.

The second half of the twentieth century saw a powerful dynamic in the U.S. economy emerge between venture capitalists and the entrepreneurs who played an instrumental role in scientific research and development (R&D). Many technologies and businesses, including Amgen, Apple, Genentech, Google, Hewlett-Packard, Intel, and Microsoft, are the fruits of venture capital.

Venture capital rose to prominence in the United States after World War II and has since become a critical ingredient in facilitating a greater payoff to R&D. Over the past fifty years, U.S. venture capitalists have raised over $600 billion, which is, on average, more than $12 billion per year. This may strike readers as a relatively small amount compared to the tens of trillions of dollars associated with general economic activity. There are some solitary investment managers on Wall Street today who are responsible for managing more money than what has been raised by U.S. venture capitalists over the past sixty years or so. But do not let the numbers mislead you—while the amount of money raised by U.S.

venture capitalist has been relatively small, its impact on the U.S. and global economies has been nothing short of spectacular.

Table I.1 provides summary statistics on the number and impact of venture capital–backed companies as a fraction of independent U.S. public companies founded after 1974. The statistics highlighted in the table speak for themselves. Venture capital–backed companies account for a whopping 85 percent of total R&D spending, nearly 40 percent of total revenues and jobs, and over 60 percent of total stock market capitalization. It is truly astonishing how such a small pool of capital can have such a large effect on economic activity.

Venture investments, by their nature, are risky. Almost all—roughly 90 percent—of startup companies fail. Venture capitalists know the highly risky nature of investing in new enterprises, but they are risk takers by nature. They put their money in companies where small investments can generate enormous returns. Venture capital is well suited to the development of entrepreneurial companies that commercialize transformative technologies.

Table I.1 Venture Capital–Backed Companies Among Public Companies Founded After 1974[*]

	Venture capital–backed	Percent	Total
Number	556	42%	1,339
Enterprise value, $b	4,136	58%	7,200
Market capitalization, $b	4,369	63%	6,938
Employees	3,083,000	38%	8,121,000
Revenue, $b	1,222	38%	3,224
Net income, $b	151	61%	247
Research and development, $b	151	85%	135
Total taxes, $b	57	59%	98

[*]All measures are as of 2014.

Source: Will Gornall and Ilya A. Stebulaev, "The Economic Impact of Venture Capital: Evidence from Public Companies," Stanford University Graduate School of Business Research Paper No. 15-55, November 2015.

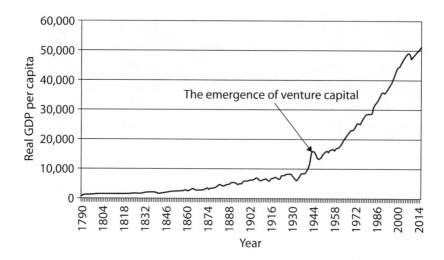

FIGURE I.I U.S. real GDP per capita (in constant 2009 dollars). *Source*: MeasuringWorth.

Companies funded by venture capital invest heavily in research and development. Venture capital–backed companies account for over 40 percent of the R&D spending by all U.S. public companies today. That spending produces value, not just for those companies, but for the entire U.S. economy, and the global economy, through what economists refer to as "positive spillovers."

The emergence and maturation of the venture capital business in the United States has coincided with a sharp rise in real gross domestic product (GDP) (figure I.1). The payoff to R&D related to venture capital has propelled living standards in the United States to new heights, fundamentally reshaping business and economic landscapes in the process.

Venture Investing and the Payoff to Research and Development

The economic prosperity of the postwar period in the United States came from several factors, spanning the public and private sectors.

Chief among these were scientific advances that produced a range of innovative new commercial products and businesses in sectors including computing and communications, medicine, transportation, energy, and consumer products.

What is the link among deep science–related R&D, technological change, and economic dynamism? In the economic literature, production, as captured in a standard Cobb–Douglas production function, measures outputs from capital inputs (K), labor inputs (L), and R&D inputs (R), indicating that output is related to the function of its inputs. Consider the relationship between economic inputs and outputs as illustrated in figure I.2.

Each input, such as R&D (R), has an elasticity component (α) related to the input it represents; for example, R^{α}. Elasticity refers to the response of a given change in output for a given change in input, typically measured in percentage terms. In the standard economics model of production, an increase in R&D (R) increases the overall output produced with a given quantity of capital and labor. The magnitude of this contribution is measured by the rate of growth multiplied by the elasticity (α) of output. The elasticity component is its payoff. The payoff component turns out to be very important for R&D's impact on output and productivity.

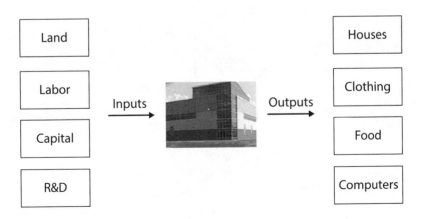

FIGURE I.2 The production process.

Historically, decreasing the payoff to R&D has influenced output and productivity. If R&D is not transferred, refined, or diffused from the lab to the market, it may have little effect on output and productivity. Without a system to realize its full benefit to commercial application, R&D may not generate significant increases in productivity or growth.[1]

The U.S. federal government, often in collaboration with academia, companies, venture capitalists, and the entrepreneurs they back, has long supported scientific research and technological development. The interplay of various public and private agents toward science-based technological development and ensuing commercialization is depicted in figure I.3.

There is no magic wand the federal government can wave to produce a payoff to R&D. The fact is that the ecosystem that produces innovation and prosperity in the United States is diverse. While all the various constituents of the ecosystem that produce a payoff to R&D are important, venture capital stands out as

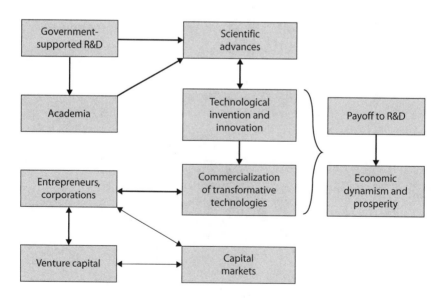

FIGURE I.3 Venture investing and the R&D payoff.

the most dynamic catalyst. Indeed, policymakers in Europe, Asia, and other parts of the world recognize the vital contribution of venture capital in fostering economic growth and prosperity. The record in the years following World War II shows venture capital playing a key role in fostering the development of promising, high-risk, transformative innovations based on scientific advances.

While venture capital has fostered the development and commercialization of transformative technologies over the past sixty years, there was a marked shifted in the venture capital business in the first two decades of the twenty-first century. This shift is a migration away from supporting transformative technologies based on deep science toward a concentrated interest in software investments. More than ever, a greater share of government-backed scientific R&D activity is receiving less attention and support from U.S. venture capitalists. This trend, discussed in chapter 4, is producing what complexity scientists refer to as a "diversity breakdown."

There are many reasons for the pronounced shift in U.S. venture investing. One of the primary concerns related to this change is the potential for a reduced payoff to science-related R&D. At present, the government is pouring billions of dollars into deep science–based R&D, whereas venture capitalists are migrating away from investing in deep science. The increased concentration of venture capital in software investments reduces the potential for a significant payoff to our government's persistent R&D efforts.

The shift away from deep science–backed venture investing over the past decade coincides with a notable deceleration in U.S. productivity and economic growth, both components of what we refer to as "economic dynamism."[2] U.S. productivity growth has been anemic over the last ten years, rising at a pace 40 percent below its historical postwar average of 2.2 percent and on a par with the dismal performance of the 1970s (figure I.4).

The anemic growth in U.S. productivity mystifies economists. Princeton University economics professor and former vice-chairman

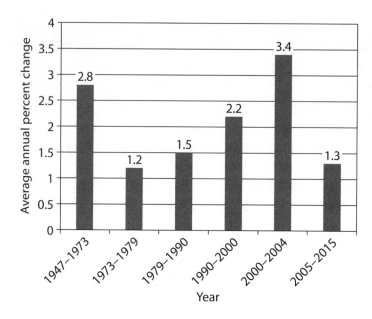

FIGURE 1.4 U.S. nonfarm business-sector productivity growth. *Source*: U.S. Bureau of Labor Statistics.

of the Federal Reserve, Alan S. Blinder, notes that unveiling the mystery behind the recent slowdown in U.S. productivity growth involves looking at technological improvement.[3]

Consider, says Blinder, the possibility that technological progress has actually slowed in recent years, despite all the whiz-bang stories one sees in the media. It is not clear at all if venture-backed investments in social media and related software applications, such as Twitter and Snapchat, facilitate more output from the same inputs of labor and capital—which is how economists define technological progress. Some online services, notes Blinder, might even reduce productivity by turning formerly productive work hours into disguised leisure or wasted time.

Two leading productivity experts, John Fernald of the Federal Reserve Bank of San Francisco and Robert Gordon of Northwestern University, argue that the greatest productivity gains from information technology came years ago and that recent technological inventions look small by comparison. In his new book, *The Rise and Fall of American Growth*, Gordon observes that economic growth is not a steady process that creates economic advances at a regular pace. He notes that economic progress can occur more rapidly in one period verses another period: "Our central thesis is that some inventions are more important than others, and that the revolutionary century after the Civil War was made possible by a unique clustering, in the late nineteenth century, of what we will call the 'Great Inventions.' "[4]

According to Gordon, the recent decline in economic growth is a result of advances now being channeled into a narrow sphere of activities focused on entertainment, communications, and the collection and processing of information. Compare Facebook with the Internet, or the Apple Watch with the personal computer. Perhaps inventiveness has not waned, but the productivity-enhancing impacts of recent technology and innovation backed by venture investors may have. There is also economic evidence, as noted by University of Maryland economist John Haltiwanger, that American businesses are now churning and reallocating less labor than they have in the past.[5] The downside of this activity is less entrepreneurial dynamism, which in turn slows down efficiency gains.

Signs of waning economic dynamism in the United States portend more challenging times ahead for businesses as well. A report by Ian Hathaway and Robert E. Litan published in 2014 by the Brookings Institution reveals a slowing in business dynamism in America.[6] Business dynamism is the process by which firms continually are born, fail, expand, and contract, as jobs are created, destroyed, and turned over. Hathaway and Litan state that the U.S. economy has become less entrepreneurial, as evidenced by a persistent decline in business churning and new firm formation. The authors point out that the rate of firm entry in the United States has declined from around 15 percent in the late 1970s to around 8 percent currently, amounting to nearly a 50 percent falloff.

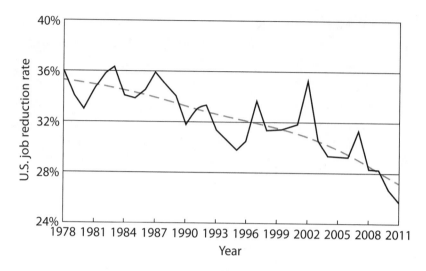

FIGURE I.5 U.S. job reduction rate and trend, 1978–2011. *Sources*: U.S. Census Bureau; Robert E. Litan and Ian Hathaway, "Declining Business Dynamism in the United States: A Look at States and Metros," Brookings, May 5, 2014; author's calculations.

Note: Trend was calculated by applying a Hodrick–Prescott filter with a multiplier of 400.

Additionally, the rate of job reallocation has been declining steadily over the past several decades (figure I.5).

The decline in business dynamism in the United States is not an isolated phenomenon. The firm entry rate and reallocation rate in nearly all key segments, including manufacturing, transportation, communications, and utilities are slowing (figure I.6).

The overall message from Hathaway and Litan's research is clear: business dynamism and entrepreneurship are experiencing a troubling long-term decline in the United States. Older and larger businesses are doing better than younger and smaller ones, yet smaller-growth firms are the predominant source of new job creation. Firms and individuals appear to be more risk averse, too—businesses are hanging on to cash, fewer people are launching firms, and workers are less likely to switch jobs or move.

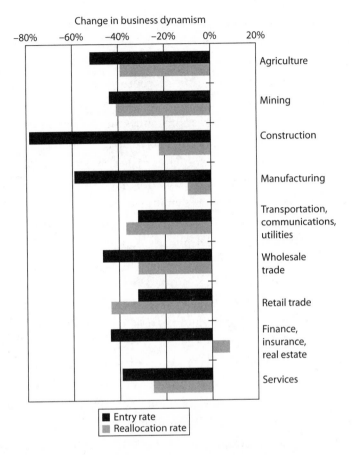

Change in business dynamism

FIGURE 1.6 Percentage change in business dynamism by sector, 1978–2011. *Source*: U.S. Census Bureau; Robert E. Litan and Ian Hathaway, "Declining Business Dynamism in the United States: A Look at States and Metros," Brookings, May 5, 2014; author's calculations.

A recent article by James Surowiecki confirms this problem in business dynamism. According to Surowiecki, research from Jorge Guzman and Scott Stern shows that while high-growth firms are being formed very actively, they are not succeeding as often as such companies once did.[7] Surowiecki surmises that this is owing to the increased power of incumbents. As Hathaway and Litan also point out, American industry has grown more concentrated over the last thirty years, and incumbents have become more powerful in almost every business sector.

The Brookings economic study does not mention the unprecedented change in the allocation of venture capital investing that is impeding the payoff to scientific R&D, nor does it discuss the significant regulatory changes and shifts in U.S. capital markets over the last couple of decades that have created strong headwinds for small, entrepreneurial enterprises. These headwinds, discussed in chapter 5, constrain the most dynamic part of the U.S. economy and work to reduce economic dynamism over time.

The productivity slowdown is a major concern, although it is premature at this juncture to draw any definitive conclusions about what it could mean, since the productivity payoff from technological invention and innovation is often not immediately observed. MIT Sloan School of Management professors Erik Brynjolfsson and Andrew McAfee note that it took decades to improve the steam engine to the point that it could fuel the Industrial Revolution.[8] It may well take the recent crop of new digital technologies years and decades to pay off in greater efficiency and prosperity.

The type of economic environment that makes the best use of new digital technologies, say Brynjolfsson and McAfee, is one that is conducive to innovation, new business formation, and economic growth. They see five broad areas of focus: education, infrastructure, immigration, entrepreneurship, and basic research. With respect to the last two, entrepreneurship is a vital catalyst for new job creation, while basic research requires the backing of government, as corporations more recently have tended to concentrate on applied research and incremental change. The fact that nondefense federal research and development spending as a percentage of GDP has declined by more than a third over the past two decades is also a concern.

Economic dynamism in the United States depends, as venture capitalist and author George Gilder notes, on the invention, investment, and distribution of venture capital, capital markets, and entrepreneurs. Venture-based companies in the past have accounted for over 20 percent of U.S. GDP and almost two-thirds (60 percent) of total U.S. stock market capitalization. As Gilder states matter-of-factly, "Venture companies are the prime source of American growth."[9] But this time seems different.

A chief concern among economists today is the rising concentration of venture investments in software, which has given rise to a diversity breakdown in venture capital. Symptomatic of this diversity breakdown is the rise of "unicorns." There are more than one hundred venture-backed companies with private valuations greater than $1 billion, with a combined market capitalization approaching $0.5 trillion dollars.

In the history of venture capital, no private company but Apple has obtained anything close to a billion-dollar market capitalization ahead of becoming a public company. Venture-backed companies such as Cisco, Intel, and Microsoft reached that level through IPOs that led to long years of building value in rising stock market capitalizations. Most venture-backed companies in the past filed IPOs with market capitalization in the millions and tens of millions of dollars. Until 2000, over 70 percent of all IPOs annually raised less than $50 million on average. Importantly, the bulk of the investment gains associated with these IPOs were accrued not by a few private institutional investors, but by millions of individual investors through public share price appreciation over time.

The ascent of the unicorns reflects a concentration of venture deals in software investments. It is a dramatic shift, and an unprecedented one. With unicorn valuations in the private market exceeding those in the public market, the traditional valuation–creation dynamic of growth—which brings with it rising stock prices and market capitalizations that benefit millions of investors and foster economic dynamism—breaks down.

The siphoning of capital away from entrepreneurs commercializing transformative technologies based on deep science, coupled with regulatory and capital market shifts, prevents investors from participating in companies that have traditionally provided the payoff to R&D and that have been a prime generator of jobs, incomes, and wealth in the United States. Yes, this time seems different in the venture-investing world, and it is not yet clear how this situation will play out in the years ahead. At the end of chapter 7, we share some of our thoughts on the future from our perspective as practicing venture investors in deep science.

Venture capital is a vital part of a healthy and dynamic economy, playing an important role in helping to generate a significant payoff to R&D. Both venture capital and entrepreneurship focused on developing and commercializing transformative technologies have proven to be crucial ingredients to increasing the payoff to R&D in the United States. Anemic productivity growth and diminished economic dynamism, should they persist, will pose significant challenges for the U.S. economy in the years ahead.

Our purpose in writing this book is to highlight the vital role venture capitalists have traditionally played with regard to R&D associated with deep science–based technologies in the hope that they will participate in it again.

Deep Science Versus Software Investing

Throughout this book, we contrast deep science investments with software investments. We do this knowing that there is not a clear and universally recognized distinction between deep science investments and software investments. For our purposes, the distinction between deep science investments and software investments is not as much a scientific differentiation as it is a business model differentiation.

To be clear, deep science investments and software investments can both intersect in their use of computer science, and we classify computer science as a deep science. Most science today is done using computers, so computer science is as critical a piece of deep science R&D in the twenty-first century.

Computer science has both a theoretical and an applied side. The theoretical side of computer science includes theories of computation, information, and coding; algorithm and data structures; and programming language theory, all of which require knowledge of deep science. The applied side includes disciplines such as artificial intelligence; computer architecture and engineering; computer security and cryptography; and concurrent, parallel, and distributed systems, which also require knowledge of deep science. The applied side also includes disciplines with more immediately recognizable applications, such as software engineering, application development, computer networking, computer graphics, and database systems.

The distinction between deep science investments and software investments in this book is not a distinction of whether software is used, but instead of which business model is appropriate to bring the products of deep science and software to market. Governments funding deep science focus primarily on technology, but investors focus primarily on business models and business model differentiation. Innovation requires business model execution and technology differentiation.

For the purposes of our writing, the difference between the deep science business model and the software business model resides in (1) the capital and resources required to bring a product to market initially, and (2) the difference in timing between the investment in resources during development and the market validation of the product. We realize these will necessarily be generalizations, but we believe the distinction remains important.

In the deep science business model, significant capital and resources must be brought to the development of the product before it can be released to the market. Often, hardware development is necessary, as equipment is needed to make the end product. Examples include semiconductor chip manufacturing, quantum computing development, electric car manufacturing, and biotechnology drug development.

In the case of developing a new semiconductor chip technology, investments to prove out the chip architecture are followed by investments to make the chip at scale and improve yield, all before the chip can be released to the market. This investment may be anywhere from $50 million to $100 million, and these funds will be needed before it can be ascertained if the market will indeed adopt this new semiconductor chip technology.

In the case of developing biotechnology drugs, the investment and timing of the investment in comparison to market adoption is even clearer. Approximately ten years and hundreds of millions of dollars are often required to move a drug candidate into the clinic and through the three phases of clinical trials. The majority of drugs fail in this process. It is only after the Food and Drug Administration has approved a drug following phase III clinical trials that a product can be released to the market, and the success of the product in the marketplace can be gauged.

In contrast to this, the software business model requires far less capital in the earliest stages, and market adoption can be gauged before significant investment is made in scaling and revenue growth. Software applications can be written in a matter of days, weeks, or months relatively inexpensively. These applications penetrate the market quickly through a large installed base of hardware devices. For example, software business models such as Facebook, Twitter, and gaming apps can be developed with software engineers and very few fixed assets. The sole costs are labor and laptops in the simplest examples. Then, owing to the Internet, it can be determined relatively quickly and inexpensively if there are users for the product. After this has been determined, only then is significant capital required to scale the business and drive large amounts of revenue.

In the deep science investing paradigm, it can take at least five to ten years and an investment of $50 million to $100 million, or more, to get a product to market. In contrast, a typical software business model may take as little as weeks and hundreds of thousands of dollars to get a first product to market. As one can imagine, the investors for each business model and the time frame over which investors are looking to invest can be very different for each of these business models.

As we will see in chapter 4, venture capital has migrated away from the deep science business model toward software business models. This migration has major economic implications, given that deep science has historically played an integral role in fostering innovation and prosperity.

Venturing into Deep Science

In the chapters ahead, we will highlight and analyze each element that results in a payoff to R&D and leads to increased economic dynamism and prosperity (as shown in figure I.3). We begin with a look at the evolution of deep science and the technologies spawned by deep science over the past three centuries and their impact on both economic and business landscapes. We follow this discussion with an in-depth discussion of the deep science innovation ecosystem and the evolution of the U.S. venture capital business. These chapters lay the foundation to discuss the significant shift in the venture investing landscape that we discuss in the final chapters of this book.

I

Deep Science Disruption

The farther back you can look, the farther forward you are likely
to see.

—WINSTON CHURCHILL

AS WE NOTED IN the introduction, there is a disconcerting trend
in venture investing casting an ominous cloud over the U.S. economy: there is less focus and funding for early-stage, entrepreneurial companies commercializing technologies based on deep
science. In this book, we use the term "deep science" to refer to
what is sometimes called "hard science." Deep science consists of
any of the natural or physical sciences, such as chemistry, biology,
physics, or astronomy, in which aspects of the universe are inves-
tigated by means of hypotheses and experimentation. These disciplines form the core of modern science. Along with mathematics,
the language frequently used by scientists to construct scientific
theories, these sciences are also the foundation of what we call
deep science.

By its nature, deep science is revolutionary. It fosters new
research and investigative methods, which in turn lead to invention and innovation. Deep science drives new product development and commercialization. Deep science alters the way atoms
and bits are configured over time. The new configurations manifest

in the form of new technologies, such as electricity and the microprocessor. These two transformative technologies are the products of advances in deep science associated with the pioneering work of Faraday, Maxwell, Planck, Einstein, Feynman, and others.

The U.S. economy and economies overseas have prospered greatly as a result of technological inventions and innovations that have their roots in deep science. The Industrial Revolution had its roots in the classical physics pioneered by Sir Isaac Newton. The modern age of digital computing and the Internet has its roots in the deep scientific advances of quantum mechanics and information theory.

Let us begin by taking a trip back to the time when the deep science links were forged so that, in the spirit of Churchill, we may see further ahead and have greater insight into the decline in U.S. economic dynamism seen today.

Deep Science Comes of Age

In 1687, the English physicist and mathematician Isaac Newton published his now famous scientific treatise, *Philosophiæ Naturalis Principia Mathematica* (*Mathematical Principles of Natural Philosophy*, or simply, *Principia*). The publication of *Principia* set in motion a powerful dynamic in science that revolutionized the way people thought about the nature of the universe. A new age of experimental science in astronomy and physics was born. These advances, in turn, inspired new technological inventions and innovations that continued to disrupt old inventions and innovations in a process of creative destruction that Schumpeter saw as the defining characteristic of the capitalist economy. Box 1.1 details the disciplines of deep science.

Newton's *Principia* provided a new language that explained the motions of heavenly bodies as well as things on Earth. Newton's science illuminated the cosmos as a great machine, and this meme

BOX 1.1
The Disciplines of Deep Science

Physics is the branch of science concerned with the laws that govern the structure of the universe and the properties of matter and energy and their interactions. Physics is divided into branches including atomic physics, nuclear physics, particle physics, solid-state physics, molecular physics, electricity and magnetism, optics, acoustics heat, thermodynamics, quantum theory, and relativity. Prior to the twentieth century, physics was known as *natural philosophy*.

Chemistry is the area of science concerned with the study of the structure and composition of different kinds of matter, the changes matter may undergo, and the phenomena that occur in the course of these changes. Chemistry is commonly divided into three main branches: organic, inorganic, and physical. Organic chemistry deals with carbon compounds. Inorganic chemistry focuses on the description, properties, reactions, and preparation of all the elements and their compounds. Physical chemistry is concerned with the quantitative explanation of chemical reactions and the measure of data required for such explanations. Ancient civilizations were familiar with certain chemical processes; for example, extracting metals from their ores and making alloys. The alchemists endeavored to turn base (nonprecious) metals into gold, and, toward the end of the seventeenth century, chemistry evolved from the techniques and insights developed during alchemical experiments.

Biology is the science of life and the study of millions of different organisms that share in that phenomenon—their physical structure, ecological functions, social interrelationships, and origins. It encompasses a range of subdisciplines, including genetics, molecular biology, botany, zoology, anatomy, and evolution. In the twenty-first century, biology is converging with other deep science disciplines to create a new subdiscipline called the life sciences.

Astronomy is the science of celestial bodies: the sun, moon, and planets; the stars and galaxies; and all other objects in the universe. It is concerned with their positions, motions, distances, physical conditions, and their origins and evolution. It is divided into fields including astrophysics, celestial mechanics, and cosmology.

Astronomy is perhaps the oldest recorded science; there are observational records of the stars from ancient Babylonia, China, Egypt, and Mexico. The universe is explored through the use of rockets, satellites, space stations, and space probes. The launch of the Hubble Space Telescope into permanent orbit in 1990 has enabled the detection of celestial phenomena seven times more distant than that possible by an Earth-based telescope.

Computer science is the youngest of the deep science fields, with modern roots stemming back to 1835 and Charles Babbage's mechanical computer. Babbage is often regarded as one of the first pioneers of computing owing to the development of his "Analytical Engine." Babbage's assistant, Ada Lovelace, is credited as the pioneer of computer programming. Computer science began to flourish in the 1940s and 1950s with the advent of programmable electronic computing machines and the development of a new field of communications science called information theory. Pioneered by Claude Shannon in 1948, information theory is a branch of applied mathematics, electrical engineering, and computer science involving the quantification of information. Computer science is undergoing another revolution today led by advances in quantum computing. The dawn of quantum computing promises incredible advances in computational ability that dwarf even the most powerful supercomputers on the planet today.

There is a consilience of various deep science disciplines underway known as **complexity science**. Complexity science is multidisciplinary in nature, bringing together scientists from a range of fields in both the hard and soft (i.e., social) sciences to explore the dynamics of complex systems ranging from ant colonies and beehives to weather and stock markets.

Source: Scientific American Science Desk Reference (Norwalk, CT: Easton, 1999).

became fertile ground for creative minds seeking to apply powerful new mathematics and science to the invention of new ways of doing things on Earth. These advances led to the Industrial Revolution.

Newton's deep science revolution inspired, directly or indirectly, the development of many new innovative technologies in the United States, the United Kingdom, and Europe. Among them were the reflecting telescope (the telescope design used by nearly all the large, research-grade telescopes in astronomy research today), the spinning jenny, the steam engine, the gas turbine, gas lighting, the cotton gin, the metal lathe, the lithograph, and the sheet papermaking machine. A veritable Cambrian explosion of new types of machinery proliferated. As these innovative technologies emerged, they profoundly altered the basic fabric of their respective economic landscapes.

The Industrial Revolution and its economic dynamism have roots in the deep scientific innovations of Newton and other scientific giants, such as Copernicus and Kepler. It was the seventeenth century, during the Age of Enlightenment, that witnessed the beginnings of the vital links between deep science, technological innovation, and economic progress. French mathematician Jean le Rond d'Alembert summed up Newton's inspiring era of scientific discovery beautifully in 1759, stating,

> Our century is called . . . the century of philosophy par excellence. . . . The discovery and application of a new method of philosophizing, the kind of enthusiasm which accompanies discoveries, a certain exaltation of ideas which the spectacle of the universe produces in us—all these causes have brought about a lively fermentation of minds, spreading through nature in all directions like a river which has burst its dams.[1]

The Newtonian revolution was just the beginning of advances in science that would have a profound impact on the development of technology and the global economy.

As profound and inspiring as Newton's deep scientific innovations were in the seventeenth century, it was not the ultimate height of scientific discovery. As the Industrial Revolution gained momentum, the eighteenth century saw the development and refinement of Newton's classical mechanics. The work of French scientist Pierre-Simon Laplace advanced physics, astronomy, mathematics, and statistics.

Laplace translated the geometric study of classical mechanics into one based on calculus, allowing a broader range of problems to be analyzed. He was also one of the main architects of the Bayesian interpretation of probability theory in statistics. In astronomy, Laplace was one of the first scientists to postulate the existence of black holes and the notion of gravitational collapse, which continue to be topics of interest to deep scientists in the twenty-first century. Laplace is remembered as one of the greatest deep scientists of all time and is sometimes referred to as the "French Newton" or "Newton of France," because of his prodigious natural mathematical faculty, unrivaled by any of his peers.

In the nineteenth century, preeminent English scientist Michael Faraday and Scottish scientist James Clerk Maxwell inaugurated the post-Newtonian age of deep science. Faraday received little formal education but is considered one of the most influential scientists in the rich history of science. Faraday and Maxwell's insights ushered in the age of electricity and a new era of chemistry.

Faraday contributed to the fields of electromagnetism and electrochemistry through his discoveries of electromagnetic induction, diamagnetism, and electrolysis. His observations of the magnetic field around a conductor that carries a direct current established the basis for the concept of the electromagnetic field in physics. His invention of electromagnetic rotary devices, like the Faraday disk (the first electric generator), provided the foundation for electric motor technology, which fundamentally transformed the machines developed during the birth of the Newtonian age. Faraday's work in chemistry included the discovery of benzene, the invention of an early form of the Bunsen burner, the system of oxidation numbers, and the popularization of scientific terms such as "anode," "cathode," "electrode," and "ion."

Maxwell summarized and expressed Faraday's deep scientific insights mathematically. A scientist with extraordinary mathematical skills that far exceeded Faraday's, Maxwell completed the classical edifice of the Newtonian universe with the development of the classical theory of electromagnetic radiation. His equations captured the brilliant insights of Faraday and others in the field of electromagnetic radiation.

The deep scientific work of Faraday, Maxwell, and others spawned a new innovative form of technology called the "dynamo." The dynamo was an electrical generation technology that produced direct current. Dynamos were the first electrical generation devices capable of delivering power for industrial use. They provided the foundation upon which other electric-power conversion devices were based, including the electric motor, the alternating-current alternator, and the rotary converter. The AC alternator evolved to become the dominant form of large-scale power generation. Upon first setting eyes on the modern dynamo at the Exposition Universelle ("World's Fair") of 1900 in Paris, historian Henry Adams saw the deep science–enabled technology as "a symbol of infinity." *The Education of Henry Adams* describes Adams's experience:

> As he grew accustomed to the great gallery of [dynamo] machines, he [Adams] began to feel the forty-foot dynamos as a moral force, much as the early Christians felt the Cross. The planet itself seemed less impressive, in its old-fashioned, deliberate, annual or daily revolution, than this huge wheel, revolving within arm's-length at some vertiginous speed, and barely murmuring—scarcely humming an audible warning to stand a hair's-breadth further for respect of power—while it would not wake the baby lying close against its frame. Before the end, one began to pray to it; inherited instinct taught the natural expression of man before silent and infinite force. Among the thousand symbols of ultimate energy the dynamo was not so human as some, but it was the most expressive.[2]

FIGURE 1.1 Brush's Dynamo, 1881. *Source*: blogs.toolbarn.com, "The History . . . and Science of the Electric Generator," http://blogs.toolbarn.com/2014/07 /tbt-the-history-and-science-of-the-electric-generator/.

It is difficult for us today to marvel at the dynamo in the way Adams and other did back in the nineteenth century. But the dynamo (figure 1.1) was a game-changing technology of incredible proportions borne out of Faraday's and Maxwell's deep science discoveries. It also became, along with subsequent electric technology innovations, a major impetus to economic development and growth.

The first hydroelectric power plant was constructed by deep science investor and entrepreneur Nikola Tesla with George Westinghouse at Niagara Falls in 1895 (figure 1.2). The establishment of the Tesla–Westinghouse power plants was a seminal event that launched electrification of the world.

Commenting on Tesla's transformative electrification technology, Charles F. Scott, professor emeritus of electrical engineering at Yale University and former president of the American Institute of Electrical Engineers, stated, "The evolution of electric power from the discovery of Faraday in 1831 to the initial great installation of the Tesla polyphase system in 1896 is undoubtedly the most tremendous event in all engineering history."[3]

FIGURE 1.2 Adams Station, powerhouses #1 and #2 and transformer building, Niagara Falls Power Company, 1895. The first major hydroelectric power plant in the world. *Source*: Tesla Memorial Society of New York.

A vast new landscape opened up with the age of electricity. Electric power became a dominant source of energy, enabling the development of new products and applications. It enabled power for new forms of lighting in homes, offices, factories, stores, streets, and stadiums.

Electricity enabled a new generation of machines and a host of assorted consumer appliances (e.g., ovens, refrigerators, washing machines, dryers) that continue to dot the global economic landscape. New technologies, such as the telegraph and telephone, used electric power to wirelessly transmit messages over short and long distances. Enterprise flourished, and major corporations were spawned (e.g., General Electric, AT&T). These companies continue to be viable enterprises today in the twenty-first century.

Looking back on the twentieth century, we can see how the science of classical electromagnetic radiation profoundly altered the economic landscape. There may come a time in the future when electricity is no longer the dominant form of power in the global economy, but it is difficult to envision life without it today. On the occasion of the one-hundredth anniversary of Maxwell's birth, legendary scientist Albert Einstein noted that Faraday and Maxwell changed the world forever: "The greatest alteration in the axiomatic basis of physics—in our conception of the structure of reality—since the foundation of theoretical physics by Newton, originated in the researches of Faraday and Maxwell on electromagnetic phenomena."[4]

With the deep scientific advances of Faraday and Maxwell, the scientific understanding of reality began to shift profoundly. The Newtonian mechanical universe, comprised of matter in motion, which had driven science and technological innovation, was now being transformed by the deep scientific insights of Faraday and Maxwell, ultimately giving birth to the Age of Electrification.

In the century that followed Faraday and Maxwell, Einstein and his fellow scientists would be deeply inspired by their work. This inspiration would help give rise to another wave of technological innovation and creative destruction associated with the birth of another revolution—the science of quantum mechanics. Quantum mechanics, also referred to as quantum physics, is another seminal development in the evolution of deep science. The first thirty years of the twentieth century witnessed a remarkable outpouring of theoretical work on quantum mechanics by a host of deep scientific geniuses, including Einstein, Max Planck, Niels Bohr, Werner Heisenberg, and Erwin Schrödinger, among others. The lasting importance of quantum mechanics is evident over one hundred years later, as it continues to be intensely studied today.[5]

Whereas Newton's deep science illuminated mathematically the machine-like motion of heavenly bodies in space and identified powerful natural forces like gravity that form the foundation of classical mechanics, quantum mechanics yielded insights into the behavior of nature at the atomic level. Building on the foundation erected by Newton and Maxwell, the behaviors of elements at the atomic level were specified and modeled mathematically in rigorous fashion.

As work progressed, quantum mechanics developed and became recognized as the most successful of all scientific theories in terms of its ability to explain the inner workings of nature. Progress occurred despite the fact that, as Nobel laureate Richard Feynman observed, "nobody understands quantum mechanics."[6] That is not to say that quantum mechanics is the be-all and end-all of science. If history is any guide, a new revolutionary science beyond quantum mechanics lies on the horizon. The great deep science inventor, Nikola Tesla, foresaw a day coming when science would begin to study nonphysical phenomena. When deep science heads in this direction, said Tesla, it will make more progress in one decade than in all previous centuries of its existence. While envisioning scientific advances in the future, one may hear Winston Churchill's proclamation, "The empires of the future are the empires of the mind."

As with the Newtonian revolution in deep science that preceded it, the insights and illuminations associated with the development of the deep science of quantum mechanics profoundly altered the way deep scientists viewed nature in the twentieth century. Quantum mechanics yielded new insights into the behavior of nature at the atomic level, beyond the classical mechanics of Newton. Quantum mechanics further extended the frontiers of deep science into the atomic realm.

Quantum mechanics provided insights into the atomic realm that, in turn, fostered the development of new testable hypotheses that could be supported or falsified through experimentation. The deep science of quantum mechanics has facilitated a tremendous and continued amount of experimentation, and the fruits of such research manifest in the development of modern innovative technologies.

Box 1.2 provides an overview of the evolution of deep science that has contributed to the evolution of economies throughout the world.

BOX 1.2
Evolution of Deep Science

1500S

Nicolaus Copernicus: heliocentric solar system

1600S

Johannes Kepler: three laws of planetary motion
Galileo Galilei: telescope
René Descartes: principle of causality
Isaac Newton: classical mechanics

1700S

Pierre-Simon Laplace: celestial mechanics, probability theory

1800S

Lord Kelvin: thermodynamics
Michael Faraday: electromagnetism and electrochemistry
James Clerk Maxwell: classical theory of electromechanical
   radiation
Ludwig Boltzmann and James Clerk Maxwell: statistical physics

1900S

Max Planck and others: quantum mechanics
Albert Einstein: relativity theory
Claude Shannon: information theory
Edward Lorenz: chaos theory
Murray Gell-Mann and others: complexity theory

Source: Christian Oestreicher, "A History of Chaos Theory," *Dialogues
in Clinical Neuroscience* 9, no. 3 (2007): 279–289; M. Mitchell Waldrop,
Complexity: The Emerging Science at the Edge of Order and Chaos
(New York: Simon and Schuster, 1992).

Looking back over the past several centuries, one is struck
by the massive amount of change in the economic landscape.
The economies of the seventeenth and eighteenth centuries bear
little resemblance to the developed economies of today. Many of

the technologies and modern conveniences we take for granted were not present in the economies of previous centuries.

There was no electricity, no telephone, no radio, no television, no airplanes, no electrical appliances or devices, no magnetic resonance imaging (MRI) or microprocessors, no digital computers or smart devices, no cancer therapies, and no Internet.

The technologies and modern conveniences many of us take for granted today, but which have profoundly reshaped the economic landscape, have their roots in advances associated with deep science. Since the seventeenth century, advances in deep science have ushered in a torrent of technological change that has raised living standards over time while recasting how people live, work, and play. In the United States, deep science technologies catalyzed what economist Robert Gordon has called the "great century of growth," the period from 1870 to 1970.[7]

Knowledge is the domain of deep science. The knowledge associated with advances in deep science becomes embedded in the DNA of new technologies that penetrate the marketplace and revitalize the economy. The rise of deep science as a major force of technological innovation and economic dynamism is relatively recent in the evolution of humanity. Humans have possessed writing skills for about six thousand years, and the printing press was developed less than six hundred years ago. It has only been within the past three hundred or so years that deep science and technological innovation have become dominant factors in driving knowledge and learning in an economic system. Reflecting upon the progression of deep science and technological change and how recently it has ascended to become a potent catalyst for economic dynamism, one could make a strong case that we are still at an early stage of technological transformation.

Since the time of Newton, science has had two primary functions: (1) to foster knowledge; and (2) to enable us to do things. Since the beginning of the seventeenth century, scientific discovery and invention have advanced at a continually increasing rate, a fact that has made the last four hundred years profoundly different from all previous ages. A study of the advances in deep science and technological innovation over the past four centuries reveals that the more disruptive the deep scientific theory, the more disruptive the technological change and the greater the impact on the economy and society at large.

We see this vital relationship at work with Newton and the Industrial Revolution, Faraday and Maxwell and the Age of Electrification, and again in the twentieth century with quantum mechanics and the remarkable technologies associated with this field. Advances in deep science are disruptive by their very nature, and this disruptive character is reflected in the technologies they inspire. And it is the disruptive nature of deep science technologies that lies at the foundation of the dynamic process of *creative destruction*—a powerful dynamic that drives growth and productivity and fosters prosperity through higher living standards over time.

Empires of the Mind: The Rise of Intellectual Property

By analyzing the development of deep science we begin to appreciate even more the vital links between revolutionary theories (and the research and experimentation that supports those theories) and the creation of intellectual property, which paves the way for the commercialization of innovative technologies. "Intellectual property" refers to creations of the mind, such as inventions; literary and artistic works; designs; and symbols, names, and images used in commerce. Commercialization serves as a catalyst for economic dynamism in the form of new jobs, new markets, expanded output, and increased productivity. This process is well appreciated by investors in deep science, even if only slightly understood by the lay investor. The commercial potential of deep science–based intellectual property is what attracts and captivates the modern-day deep science investor. The owners of intellectual property—whether inventors, entrepreneurs, or companies—stand to benefit economically and financially from the commercialization of the ideas that underlie inventions and innovations. Consumers also benefit from the use of commercialized inventions and innovation.

President George Washington signed into law a new U.S. patent statute in 1790. The first patent was granted to inventor Samuel Hopkins for an improvement "in the making of Pot ash and Pearl ash by a new Apparatus and Process."[8] The second was issued to inventor Oliver Evans for flour-milling machinery. These patents would be the first of millions to come as science evolved. The importance of new technological inventions was recognized by the founding fathers and written into the U.S. constitution. Article 1, section 8, of the U.S. Constitution states and authorizes U.S. patent law: "The Congress shall have power . . . to promote the progress of science and useful arts, by securing for limited times to authors and investors the exclusive right to their respective writings and discoveries."

Deep science increasingly became fertile ground for the development of new intellectual property, as evidenced by the remarkable growth of patent activity in the United States over the late eighteenth century. Patent activity in the United States increased significantly in the nineteenth century, along with the deep science discoveries of electromagnetic radiation and the development of electricity and electrical machinery. During that century, nearly 650,000 utility patents were granted to inventors in the United States alone, many of which had their roots in deep science.[9]

One can imagine the astonishment at some of the deep science–related patent applications processed by the U.S. Patent and Trademark Office.[10] One can envision a nineteenth-century patent clerk reviewing a deep science–related invention application and thinking, "This patent application claims to do what?"

Patent activity accelerated in the twentieth century with the development of quantum mechanics and the innovations based on it—some of which were truly transformative and reshaped

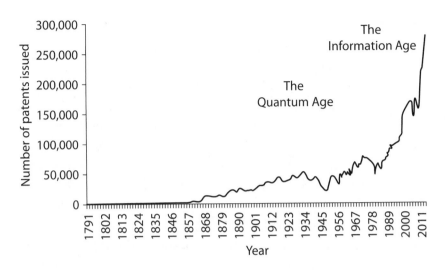

FIGURE 1.3 The number of U.S. utility patents granted between 1790 and 2013. *Source*: U.S. Patent and Trademark Office.

the U.S. and global economic landscapes. Deep science quantum–based innovations, such as the integrated circuit, microprocessor, laser, and MRI machine, likely would have astonished the great deep-scientific genius of Newton, Laplace, Faraday, and Maxwell. Advances in quantum mechanics helped fuel a remarkable increase in inventive activity, as demonstrated by the astonishing eightfold increase in patents granted, totaling over 5.3 million in the United States. Figure 1.3 traces the evolution of patent activity in the United States since the late eighteenth century, highlighting the extraordinary ascent of intellectual property during a time of great advances in deep science.

GENERAL-PURPOSE TECHNOLOGIES

In reviewing the progression of deep science over the past four centuries, something of great import emerges from the analysis: how advances in deep science manifest themselves in technology, and how technological innovation in turn impacts economic growth.

Advances in deep science manifest in technology over time, and, as they do, the structure of the economy shifts in unanticipated ways. Some deep science technologies are highly disruptive in nature. Economists refer to such innovations as "general-purpose technologies" (GPTs). General-purpose technologies unleash a powerful dynamic into the economy that persists for decades.

A GPT has the potential for pervasive use in a wide range of sectors, including energy, transportation, communications, and computing, in ways that drastically change their modes of operation.[11] General-purpose technologies are transformative and have a lasting impact on the economy. GPTs in the nineteenth century included the steam locomotive, electric dynamo, telegraph, internal combustion engine, and telephone. All had pervasive use in a wide range of sectors and significantly altered their modes of operation.

Railroads proliferated with steam locomotive technology, an innovation that changed the nature of transportation and how goods were shipped around the world. The electric dynamo gave us electric power and lighting, thus laying the foundation for modern consumer electronics in the twentieth century. The telegraph and telephone allowed information to be communicated around the world nearly instantaneously. It is difficult to understate the effect of these communication technologies at the time. They were truly revolutionary. The internal combustion engine became the technology that would power the automobile, displacing the horse and buggy as the main form of personal transportation.

Looking back, we can see the dramatic economic effect that the proliferation of deep science technologies had on the U.S. economy. The percentage of U.S. households with a car, telephone, television, radio, various machines (e.g., washing machine, refrigerator), electric lighting, and central heating—all considered modern conveniences today—rose sharply during the twentieth century (figure 1.4).

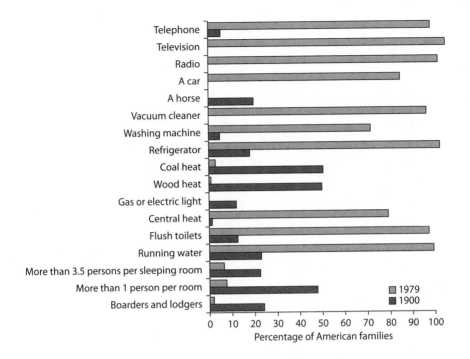

FIGURE 1.4 The penetration of new technologies in the U.S. economy since 1900, as illustrated by the change in consumption patterns of American families between 1900 and 1979. *Source*: Stanley Lebergott, *The Americans: An Economic Record* (New York: Norton, 1984), 492.

What is particularly remarkable about the nineteenth century from an economic perspective is the number of GPTs that arose. One GPT alone can have a dramatic impact on economic dynamism. The fact that many GPTs were developed during this period goes a long way toward explaining the substantial increase in economic growth and living standards also seen in that century. In the twentieth century, GPTs emerged in the form of quantum-based technologies, such as the microprocessor, personal computer, and nanotechnology. General-purpose technologies fundamentally altered the way people live, and deep science GPTs are powerful agents of economic dynamism with effects that persist for years and even decades.

LINKING DEEP SCIENCE TO TECHNOLOGICAL
CHANGE AND THE ECONOMY

The relationship between advances in deep science and the technologies that manifest from such advances is dynamic. This flexibility is one based on feedback, where developments in science lead to new innovative technologies that inspire new scientific thinking and lead to additional technological change. The deep science that led to the machine age that spawned the steam engine led to advances in thermodynamics that propelled science further.

The constant interplay between science and technological innovation, and the feedback between them, is a central element of the capitalist process that breathes new life into the economy and drives living standards ever higher. This dynamic lies at the foundation of knowledge and learning in the capitalist economy. It is what differentiates the economy associated with the Neanderthals from today's economy. The "cavemen" had access to the very same resources that entrepreneurs have access to today. What cavemen did not have back then was learning in the form of accumulated deep science and technical knowledge.

Technological change is a fact of life in a capitalist economic system. And the pace of technological change is determined within that system by scientific advances and the activities of entrepreneurs and deep science venture investors. In the early 1940s, Joseph Schumpeter made his fellow economists aware of this fact when he noted, "The fundamental impulse that sets and keeps the capitalist engine in motion comes from the new consumers' goods, the new methods of production or transportation, the new markets, the new forms of industrial organization that capitalist enterprise creates."[12]

Schumpeter rightly viewed the capitalist economy in which deep science–based technological innovations flourish as an organic process that, at its foundation, is dynamic. Technological change is a critical component of a capitalist economic system. It is not exogenous to this system, as economists often assume. The agents within the system use their respective creative talents to invent and innovate. Science provides a foundation for creativity within the system, and this creativity finds expression in new technology. Schumpeter reminded his fellow economists that the essential point to grasp is that, in dealing with capitalism, one is dealing with an evolutionary process.

As discussed by Eric Beinhocker in *The Origin of Wealth*, this model is in direct contrast to many of the models of classical economic theory.[13] Beinhocker notes that economics has historically been concerned with how wealth was created and allocated. Between the time of Adam Smith and the mid-twentieth century, an era dominated by the economic thought of Paul Samuelson and Kenneth Arrow, the first question was largely overshadowed by the second. The historical models of the economists Walras, Jevons, and Pareto began with the assumptions that an economy already existed, that producers had resources, and that consumers owned various commodities. The problem then was how to allocate existing finite wealth in the economy. As Beinhocker describes, an important reason for this focus on the allocation of finite resources is that the mathematical equations of equilibrium imported from physics were ideal for answering the allocation question, but it was more difficult to apply them to *growth*.

Beinhocker notes that the focus of conventional economic analysis on allocation rather than growth puts classical and neoclassical economic theory on shaky ground. He suggests that the complex adaptive behavior of biological systems is a more appropriate approach to economic analysis than the classical physics equations used by Walras and Jevons: "Complexity economics is a better approximation of economic reality than traditional economics, just as Einstein's relativity is a better approximation of physical reality than Newton's laws."[14]

Because classical economists have technological change exogenous to their models instead of endogenous, they have never been able to accurately foresee the economic effects of advances in deep science. There was simply no way to account for such a relationship between science and technology on the one hand and economic dynamism on the other. All economists could do was stand by and be astonished by something they had encountered but that did not fit into their economic models.

The ascent of living standards to greater heights defied the doomsday economic prophesies of economists like Thomas Robert Malthus and would continue doing so in the following centuries. In 1798, Malthus predicted food shortages owing to exponential growth in the population that would cripple economic activity. What Malthus and other economists did not factor into their long-term forecasts was a heightened pace of technological innovation associated with advances in deep science. Since the seventeenth century, deep science technological innovation has emerged as a powerful dynamic, rendering the science of accurate long-term economic prediction highly problematic, because we cannot foresee what technologies will come to pass, and how they may improve living standards over time.

Economists—even in this day of mathematical and computer sophistication—have no reliable way to predict the timing of advances in deep science, nor can they predict how such advances will impact the future pace of technological innovation. Deep science venture investor and author George Gilder notes that the glaring deficiency in traditional economics has long been its inability to explain the scale of per capita economic growth over the last several centuries. Gilder points out that the leading economic growth model explains only 20 percent of the nearly 120-fold expansion in output since the end of the eighteenth century.[15]

What traditional economics has not accounted for is the phenomenal increase in knowledge that has accompanied advances in deep science. Unlike traditional economic inputs, such as land, labor, and capital, there are no inherent limits to knowledge. In his book *Knowledge and the Wealth of Nations*, economic journalist

David Warsh provides evidence that capitalist growth has in fact been a *thousand times* greater than is registered in the conventional data.[16]

In his book, Warsh presents the findings of economist William Nordhaus in a paper titled "Do Real-Output and Real-Wage Measures Capture Reality? The History of Lighting Suggests Not." Nordhaus's analysis reveals a truly astonishing decline in the labor cost of light that accompanied advances in innovation and technology associated with deep science. Because the arrival of electricity in the 1880s resulted in a thousandfold drop in the labor cost of lighting, the advances in deep science linked to the process of technological innovation through the development of electricity led to a stunning millionfold increase in the abundance and affordability of lighting. Such is the power of the dynamic associated with deep science–enabled technologies. In the twentieth century, a similar economic dynamic would emerge with the development of quantum mechanics, microprocessors, and digital computers.

In studying the history of lighting, Nordhaus concludes that economists tend to be blind to the accumulation of knowledge and learning that occurs over time. His analysis reveals that virtually all economic growth is driven by entrepreneurial and scientific creativity. Advances in deep science fuel the accumulation of knowledge. New scientific insights manifest over time in new technologies. These new deep science–enabled technologies, whether electricity and the dynamo or the microprocessor and digital computer, foster economic growth through entrepreneurial activity over time.

The deep science–technological change dynamic is very difficult, if not impossible, to predict over time with current economic models. An attempt to generate accurate predictions with econometric models is largely an exercise in futility precisely because of the unpredictability of advances in deep science. In the presence of this unpredictability, forecasters often resort to extrapolating current trends into the future. This is a common forecasting technique, but the results are rarely satisfactory and often look silly in hindsight.[17]

Deep Science–Driven Economic Dynamism

Advances in deep science, from Newton and classical mechanics in the sixteenth century to Feynman and quantum mechanics in the twentieth century, are the foundation of an astonishing and unprecedented rise in living standards. The Industrial Revolution that followed in the wake of the Newtonian revolution in science ushered in the age of machines, which profoundly altered the economic landscape. The proliferation of machinery led to massive increases in productivity, which in turn propelled living standards upward over the course of many decades. New deep science–based technological inventions and innovations fostered the development of new businesses, industries, and ways of living. Real gross domestic product (GDP) per capita, a proxy for living standards, in the United States in the seventeenth century was less than $500. By the end of the nineteenth century, it had risen some twelvefold to $6,000. In hindsight, and with the benefit of the long view, what we can see today is the manifestation of technologies associated with advances in deep science propelling living standards to greater and greater heights. Never before had there been such a profound economic shift as that marked by the Newtonian scientific revolution.

By noting the progression of living standards over time, as reflected in figure 1.5, we can begin to appreciate the economic impact of deep science and the role investment plays in fostering the commercialization of transformative technologies based on deep science. Living standards are a general measure of economic prosperity and are highly correlated with productivity gains that result from new technological inventions and innovations. In analyzing the relationship between deep science and economic dynamism over the past four centuries, we observe new technologies

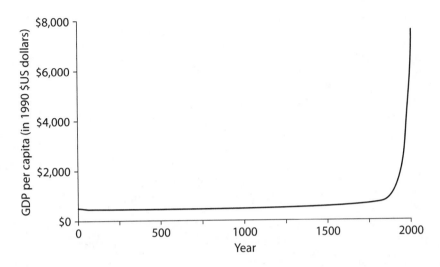

FIGURE I.5 World GDP per capita: the long view. *Source*: Adapted from Angus Maddison, in Louis D. Johnston, "Stagnation or Exponential Growth—Considering Two Economic Futures," MinnPost, November 24, 2014, https://www.minnpost .com/macro-micro-minnesota/2014/11/stagnation-or-exponential-growth -considering-two-economic-futures.

associated with deep science as driving forces of productivity growth. The development of new deep science technologies has a pronounced impact on the efficiency of an economic system.

The deep science technologies of the eighteenth century became the foundation for further technologies developed during the Industrial Revolution. With new technologies came new businesses, jobs, products, and related services. Creativity and business flourished like never before as inventors joined forces with entrepreneurs and investors to create new enterprises. It is during this period that living standards in Europe and the United States, as measured by GDP per capita, began to increase appreciably, a testimony to the economic dynamism associated with deep science–driven technological innovation. The dramatic shift in living standards, as indicated by the sharply upward-sloping line in figure 1.5, is the single biggest economic development of the past four hundred years. It has no precedent.

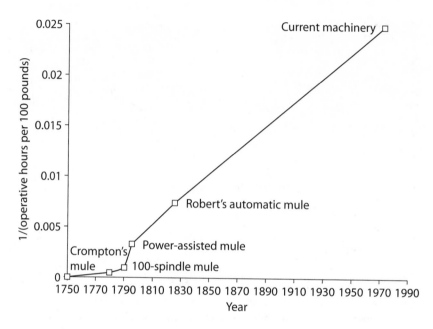

FIGURE 1.6 Productivity growth in spinning cotton thread. *Source*: Harold Catling, *The Spinning Mule* (Newton Abbot, UK: David & Charles, 1970).

The proliferation of machinery on the economic landscape had a major impact on productivity growth. Productivity growth in spinning cotton thread, for example, has risen sharply since 1750 (figure 1.6). Prior to the Industrial Revolution, it took fifty thousand hours or more to convert one hundred pounds of raw cotton into thread by hand. The earliest spinning machines at the end of the eighteenth century accomplished the same task in two thousand hours, amounting to a whopping twenty-five-fold increase in productivity. A study done in the late nineteenth century of 672 industrial processes recently converted from hand to machine production found similar dramatic gains in productivity from mechanization.

For example, the switch from hand- to machine-weaving of 36-inch-wide gingham cloth reduced the man-hours needed to produce 500 yards from 5,039 to 64, a seventy-nine-fold increase in labor productivity. But this gain was not out of the ordinary. Machine printing and binding was 212 times more efficient than hand production, whereas the machine productivity of screws was a whopping 4,032 times more efficient than hand production.[18]

We saw earlier that Newton's deep science set in motion powerful economic forces that culminated in the Industrial Revolution. Prior to Newton, economic output was mainly a function of labor and land. Generating additional output from a given parcel of land required greater labor input. One could imagine some refinements in how labor went about its various economic activities to enable increases in labor productivity (e.g., employing more oxen to assist with various agricultural activities). On the whole, however, such adjustments would lead only to modest increases in efficiency and living standards.

The economic landscape shifted dramatically in the years following Newton's revolution. New forms of deep science–inspired nonelectrical machinery proliferated, which had a significant impact on productivity over time. With the presence of new machinery came the ability to produce products more efficiently; these were not only existing products, but also new products, markets, and jobs. The machines of the eighteenth century spawned the Industrial Revolution. The Industrial Revolution, in turn, had a pronounced effect on living standards, as indicated by the dramatic increases in productivity, which, in turn, boosted GDP per capita.

Deep science–based technological invention and innovation were pervasive in the United States and overseas during this period. The invention of the electric dynamo was a seminal event in the annals of deep science–based technologies. The direct-current electric machine, similar to the Gramme dynamo, inspired deep scientist Nikola Tesla's alternating-current machinery.[19] Tesla developed and refined his alternating-current machines in the late eighteenth century, and they proliferated greatly in the early twentieth century through the rise of central stations that distributed power to large segments of the population.

The economic effects of technological change would become even more pronounced in the nineteenth century with further advances in deep science and the development of electricity and electrical machinery. The arrival of electricity in the 1880s, for example, produced a thousandfold drop in the cost of lighting, unleashing one of the most astonishing increases in economic dynamism in the history of humankind (which meant a million-fold increase in the abundance and affordability of light itself).

As breathtaking as the eighteenth century was from a techno-logical perspective, the nineteenth century was even more so. A host of game-changing deep science–based technologies prolifer-ated in the U.S. economy and economies overseas. Through this process, technology fundamentally altered the way people lived, the way they were educated, the way they worked, the businesses they created, and the enterprises and companies they worked in.

Since the early nineteenth century, productivity—defined as out-put per unit of input—in the U.S. economy as a whole has doubled about every seventy-five years, whereas labor productivity growth has been somewhat more rapid. These gains have been crucial for living standards and overall well-being. New ideas, particularly those resulting in new mechanical and electrical devices, have resulted in some extraordinary changes in productivity, which, in turn, have driven GDP per capita higher over time.

By the late nineteenth century, hundreds of industrial processes were converted from hand to machine production, resulting in impressive gains in productivity. These gains, in turn, led to ris-ing living standards in the United States and around the world.[20] Significant efficiency gains associated with the proliferation of machinery in the eighteenth century led to a near doubling in liv-ing standards in U.S. real GDP per capita in the United States, from less than $1,000 in the early eighteenth century to nearly $2,000 by the third decade of the nineteenth century.[21]

Living standards continued to soar in the wake of advances in deep science associated with electrification and the work of

Faraday, Maxwell, and others. The proliferation of electric machinery and lighting powered a doubling in U.S. living standards in the late nineteenth century over a relatively short period of time. Real GDP per capita rose from $3,040 in 1870 to $6,004 by the beginning of the twentieth century.

Prosperity, as reflected in sharply rising living standards, related to advances in deep science would continue into the twentieth century, led by the development of a new branch of physics known as quantum mechanics. During that time, U.S. living standards rose another 130 percent amid the increasing penetration of electrical machines, lighting, and other machinery and the new businesses and industries related to them.

The birth of quantum mechanics at the end of the nineteenth century and its subsequent development in the twentieth century ushered in a revolution in transformative technologies on par with the revolutions spawned by Newtonian classical mechanics and Faraday electrification. The deep science of quantum mechanics would begin to manifest in the form of new technologies in the late 1940s. These new technologies would profoundly reshape the economic landscape in the United States and overseas.

The science of quantum physics has helped foster the development of an astonishing range of technologies, from semiconductors and microprocessors in computer chips to lasers in communications systems and consumer products, MRI machines in hospitals, and much more. The proliferation of these transformative deep science technologies ushered in a level of prosperity never before experienced. During the second half of the twentieth century as quantum technologies proliferated, real GDP per capita in the United States experienced a threefold increase, from $14,400 in 1950 to $43,200 in 1999.

Inventions associated with quantum mechanics account for an estimated one-third of U.S. GDP today.[22]

Quantum technologies have created trillions of dollars of new wealth over the past century, as reflected in the U.S. stock market (figure 1.7). Quantum physics has been foundational for the

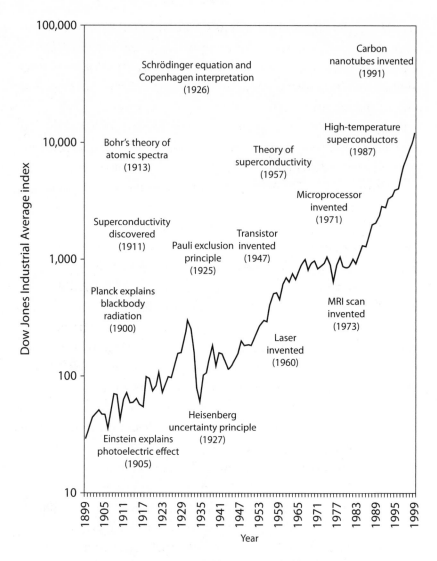

FIGURE 1.7 The ascent of the quantum economy. *Source*: Dow Jones and Company.

development of new products and industries that support tens of millions of jobs. And, while quantum mechanics is a deep science that is over a century old, its contribution to annual output in the United States is likely to continue rising in the years and decades ahead as new quantum-based technologies permeate the market. On the horizon lie transformative technologies such as quantum computers, carbon-nanotube memory devices, and a host of other quantum technologies that will be important in a variety of economic sectors, including communications, energy, transportation, and medicine. These technologies have the potential to stimulate economic dynamism and lift living standards to new heights.

In reviewing the progression of GDP per capita over the long sweep of history, we encounter a truly remarkable and astonishing rise from the eighteenth century, following the Newtonian deep science revolution, through the twentieth century. One might have presumed that the sevenfold rise in world population since 1800 should have tempered growth in living standards. However, the rise in population to nearly six billion by the end of the twentieth century pales in comparison to the growth in output as measured by GDP over the same period. The growth in output has far outstripped the growth in population, increasing nearly 120-fold. This spectacular absolute increase in GDP has driven living standards to levels today that exceed those of eighteenth-century kings and queens.

The cumulative rise in living standards in the United States since the eighteenth century has been staggering, as indicated by a fortyfold increase in real GDP per capita between 1790 and 2000 (figure 1.8).

Looking back at the progression of GDP per capita since the seventeenth century in relation to the living standards of preceding centuries, one may be forgiven for thinking that the global economy blasted off in a rocket ship with Isaac Newton and his fellow deep scientists. The rise in GDP per capita is unprecedented in previous economic history.

It was during the quantum revolution that emerged in the 1940s that the business of venture capital in the United States was in its early stage of development. Venture capital played a vital role in seeding and nurturing new entrepreneurial companies developing

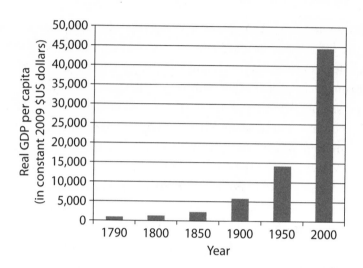

FIGURE 1.8 U.S. real GDP per capita, 1790–2000. *Source*: MeasuringWorth.

transformative technologies based on deep science. The combination of deep science and venture capital proved to be potent as a force of economic dynamism in the United States during the second half of the twentieth century.

Venture capitalists possess a unique skill set that helps foster the development and commercialization of deep science technologies. The genesis of the modern-day institutional venture capital business in the United States is intimately woven with the evolution of deep science as discussed in chapter 3. Technological possibilities are an uncharted sea, as Schumpeter noted back in the 1940s during the early days of the U.S. venture capital business. It is, in fact, this uncharted sea of technological potential that venture investors in deep science seek to capitalize. As we will see in chapter 3, U.S. venture capital has its roots in fostering technologies associated with advances in deep science.

The evolution of deep science in the twentieth century also spawned the development of a diverse ecosystem of agents including government agencies, academic researchers, nonprofit organizations, corporations, entrepreneurs, venture capitalists, and other institutional and retail investors. Collectively, these agents help facilitates the research, development, and commercialization of transformative technologies that, in turn, stimulate economic dynamism and fosters efficiency gains that fuels prosperity. They represent what we call the "Deep Science Innovation Ecosystem." In the next two chapters, we will see how important this ecosystem has become for economic dynamism over the past seventy years.

2

The U.S. Deep Science
Innovation Ecosystem

AS WE SAW IN the previous chapter, scientific discoveries and innovation are crucial elements of a dynamic economy. In order to realize the benefits of growing productivity, a prosperous nation requires research and development (R&D) in deep science followed by innovation that diffuses the resulting technologies throughout the economy. The importance of these ingredients has grown over time as domestic and global economies have developed, and policymakers in the United States and around the world recognize this importance. Such innovation improves living standards over time and provides benefits such as local job creation, improved health care and safety, new company formation, and national security enhancement.[1]

The ensuing payoff to deep science R&D efforts is accomplished through innovation. Corporations, entrepreneurs, angel investors, venture capitalists, and even public markets play a part in this process, and this array of agents helps create goods and services and, in the process, jobs and economic dynamism.

Corporations, as well as venture capitalists and entrepreneurs they support, play key roles in the innovation process. The products and related services that have emerged from deep science technologies with financial support from corporations and backing from venture capitalists have created new jobs and driven income, wealth, and living standards to ever-higher levels over time. The emergence of venture capital in the last half of the twentieth century was a major boon to the growing economic dynamism in the United States. It fostered the commercialization of deep science technologies and related services, including telecommunications, health care, cloud computing, global positioning systems (GPS), and automation.

The investment activities associated with venture capitalists are vital for economic dynamism, but venture capitalists do not work in a vacuum. Before we can truly appreciate the role of venture capital in fostering economic dynamism, we must first look at the ecosystem that constitutes the foundation of deep science innovation in the United States. This ecosystem includes corporate and government funding of R&D, as well as venture capital. It is from this foundation that ideas and technologies emerge, and it is these ideas and technologies that drive innovation. We will begin this chapter with a focus on the funding ecosystem for R&D in the United States and conclude by introducing the entrepreneurs and venture capital investors who translate R&D discoveries into innovation.

Funding R&D Science in the United States

The deep science innovation ecosystem is integral to the commercialization of deep science technologies. When venture capitalists and entrepreneurs collaborate in deep science investment activities, they become part of this larger dynamic. It involves many agents,

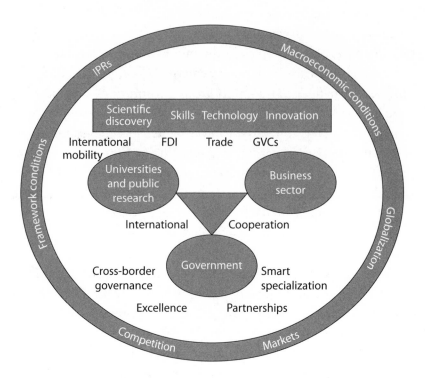

FIGURE 2.1 The deep science innovation ecosystem: foreign direct investment (FDI), global value chains (GVC), and intellectual property rights (IPR). *Source*: Organisation for Economic Co-operation and Development, *OECD Science, Technology and Industry Outlook 2014* (Paris: OECD Publishing, 2014).

including private investors (e.g., individuals, trusts, family offices), corporations, government agencies, academia, and charitable or nonprofit organizations. Some of the major elements of the deep science innovation ecosystem are depicted in figure 2.1.

The corporate sector accounts for the lion's share of the nearly $500 billion spent annually in the United States on R&D (figure 2.2, table 2.1). The other key R&D contributor is the federal government, spending a little over one-third of what is spent by the corporate sector. The remaining contributors are charitable organizations and academia, which collectively account for approximately 6 percent of annual R&D spending. The federal government spends approximately 30 percent of its R&D budget

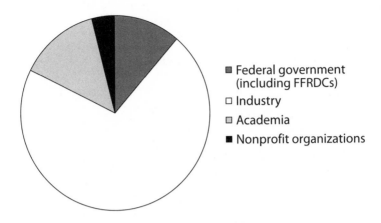

- Federal government (including FFRDCs)
- □ Industry
- ▫ Academia
- Nonprofit organizations

FIGURE 2.2 U.S. R&D spending by segment, 2014. FFRDCs are federally funded research and development centers. *Source*: 2014 Global R&D Funding Forecast, Battelle.org; *R&D Magazine*.

at universities and the vast majority of the remainder at federal labs and through its agencies.[2]

Massachusetts Institute of Technology (MIT) scientist Vannevar Bush's vision for the role of government in science has served as a viable foundation for the United States in the postwar era. Today, the U.S. federal government invests over $120 billion in R&D annually. Federal investment in basic and applied research accounts for around half this amount, with the other half going into technological development.

The federal government has worked closely with universities to foster the commercialization of deep science technologies in the United States. Bush aptly expressed the recognition of the need for government-sponsored science in 1945: "There must be a stream of new scientific knowledge to turn the wheels of private and public enterprise."[3]

Academia (figure 2.3) is the second largest R&D performer after the corporate sector. American research universities perform more than $60 billion worth of R&D annually, and nearly 60 percent of academic R&D funding is sourced from the federal government.

Table 2.1 U.S. R&D Spending in 2014, in Billions of $US Dollars (Percent Change from 2013)

	Federal government	FFRDCs (Government)	Industry	Academia	Nonprofit organizations	Total
Federal government	35.7 (1.0%)	16.5 (1.1%)	27.8 (1.1%)	37.1 (2.5%)	6.0 (1.1%)	123 (1.5%)
Industry		0.3 (0.7%)	302.5 (4.1%)	3.3 (1.7%)	1.4 (0.5%)	307.5 (4.0%)
Academia		0.1 (0.1%)		13.2 (2.0%)		13.3 (1.9%)
Other government funding		0.0 (0.1%)		4.0 (1.1%)		4.0 (1.0%)
Nonprofit organizations		0.1 (0.2%)		5.3 (2.2%)	11.3 (4.0%)	16.7 (3.4%)
Total	35.7 (1.0%)	17.0 (1.0%)	330.3 (3.8%)	62.9 (2.2%)	18.7 (2.7%)	464.5 (3.2%)

Note: FFRDCs are federally funded research and development centers.

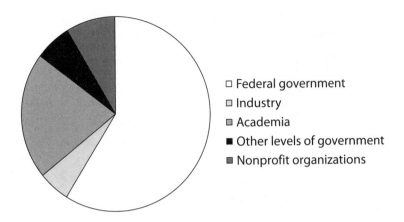

FIGURE 2.3 U.S. R&D spending in academia, 2014. *Source*: 2014 Global R&D Funding Forecast, Battelle.org; *R&D Magazine*.

Thus, we can see the close relationship between the federal government and academia associated with scientific research.

As we dive deeper into venture capital trends, it is clear that academic R&D and venture capital investing have begun to demonstrate different patterns of investment over the past decade. What is particularly noteworthy about academic R&D is a pattern of investment in the deep sciences. Over the past twenty years, the distribution of academic R&D has favored the deep sciences, especially the life sciences and physical sciences. Currently, the life sciences represent approximately 57 percent of academic R&D, whereas the physical sciences represent 30 percent of academic R&D. By contrast, information technology and the soft sciences together represent only 13 percent of academic research (figure 2.4).[4]

Nonprofit organizations are the remaining source of R&D funding for deep science in the United States. Such organizations account for around 4 percent of total U.S. research spending. Included in this group are organizations like the Carnegie Institution for Science, the Rockefeller Foundation, and the Bill and Melinda Gates Foundation. The existence of these organizations reflects previous entrepreneurial success and a mission to increase

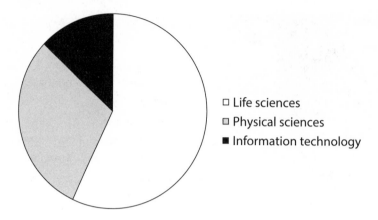

FIGURE 2.4 U.S. R&D spending in academia by segment. *Source*: 2014 Global R&D Funding Forecast, Battelle.org; *R&D Magazine*.

quality of life through philanthropic activities. A host of prominent nonprofit organizations are dedicated to scientific discovery and the support of exceptional independent research that fosters innovation. As such, nonprofit R&D activity complements the research and development work carried out by the federal government and within academia.

The nexus of governmental agencies working with university researchers and entrepreneurs on next-generation science technology is responsible for the realization of technologies that consumers, businesses, and governments take for granted today, such as computing, the Internet, mobile communication, improved health care, security, and national defense. However, there is an ongoing shift in economic resources in the United States today that is creating increasingly strong headwinds for entrepreneurs and venture capitalists developing and commercializing deep science technologies.

Government funding for scientific research is coming under pressure in the United States. Federal government research investment, as a percentage of the total federal budget, has decreased significantly over the past several decades, from 10 percent in the late

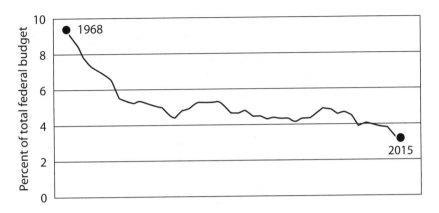

FIGURE 2.5 Federal R&D outlays as a share of total federal budget, 1968–2015. U.S. federal R&D spending is on the wane. *Source*: American Association for the Advancement of Science.

1960s to below 4 percent today (figure 2.5). This decline comes while total mandatory government spending on items such as social security, unemployment and labor, and health care (including the Medicare program) have taken an increasingly larger percentage of the federal budget. Averaging about 10.2 percent of GDP since 1973, mandatory spending is projected to increase to about 14 percent of GDP by 2023.[5]

A report published by MIT in April 2015, entitled "The Future Postponed: Why Declining Investment in Basic Research Threatens a U.S. Innovation Deficit," notes that declining U.S. federal government research investment is problematic.[6] It puts forth recommendations to expand research in the United States, including:

- Invest in research in neurobiology, brain chemistry, and the science of aging to develop new treatments for Alzheimer's disease.
- New antibiotics could tackle the growing health threat posed by the rise of antibiotic-resistant bacteria, an area where commercial incentives to invest are lacking.

- Synthetic biology research could lead to such developments as customized treatments for genetic diseases, viruses engineered to identify and kill cancer cells, and climate-friendly fuels. However, a lack of investment in laboratory facilities is leading to a migration of top talent and research leadership overseas.
- The United States has the potential to take a leadership role in a number of areas, including fusion energy research, robotics, and quantum information technologies.

Corporate Influence in Deep Science Innovation

Corporate involvement in deep science innovation in the United States has a long and distinguished lineage. Some of the biggest American corporations today have their roots in the deep science that flourished in the nineteenth and early twentieth centuries. General Electric was established in 1892 to continue the commercialization of Thomas Edison's discoveries in electricity, including electric motors, dynamos, and lighting. AT&T traces its roots back to 1879 and the invention of the telephone by Alexander Graham Bell. DuPont, established in 1802, was founded on the innovation of improved explosive powders and the commercialization of dynamite in 1880, which flourished with the building of the railroads.

Corporations were among the primary venture investors in the United States before venture investing became institutionalized following World War II. In the 1970s, corporations overtook the federal government as the largest sponsor of R&D. Since then, corporate R&D spending has ranged between 2.5 percent and 3 percent of GDP.

This increase in R&D spending by corporations was encouraged by the proliferation of products and services with roots in scientific innovation. As such, deep scientists became an integral part of corporate America in the latter half of the nineteenth century and throughout the twentieth century. R&D labs were routinely stocked with some of the best and brightest scientists tasked with helping to push the frontiers of innovation. There are scores of books and thousands of stories about the key roles played by deep scientists in various U.S. corporations.[7] Box 2.1 describes the contributions of two pioneering corporate inventors.

BOX 2.1
Deep Science Corporate Investment Pioneers

KATHERINE BLODGETT

Scientist and inventor Katharine Blodgett was educated at Bryn Mawr College and the University of Chicago. She became a pioneer in several respects: She was the first woman to receive a Ph.D. in physics from the University of Cambridge and the first woman hired by General Electric. During World War II, Blodgett contributed important research to military needs like gas masks, smoke screens, and a new technique for de-icing airplane wings. Her work in chemistry, specifically in surfaces at the molecular level, resulted in her most influential invention: nonreflective glass. Her "invisible" glass was initially used for lenses in cameras and movie projectors, but it also had military applications, such as in wartime submarine periscopes. Today, nonreflective glass is still essential for eyeglasses, car windshields, and computer screens.

STEPHANIE KWOLEK

Shortly after graduating from Carnegie Mellon University in Pittsburgh, Stephanie Kwolek began working at the chemical company DuPont, where she would spend forty years of her career. She was assigned to work on formulating new synthetic fibers, and in 1965 she made an especially important discovery. While working with a liquid crystal solution of large molecules called polymers, she created an unusually lightweight and durable new fiber. This material was later developed by DuPont into Kevlar, a tough yet versatile synthetic used in everything from military helmets and bulletproof vests to work gloves, sports equipment, fiber-optic cables, and building materials. Kwolek was awarded the National Medal of Technology for her research on synthetic fibers and was inducted into the National Inventors Hall of Fame in 1994.

The success of corporate R&D has become a priority among executives as companies seek to foster innovation to stay competitive and profitable in the marketplace during a time of accelerating technological change. Although research and development are closely related, they too often get lumped together when in fact they are distinct areas. Looking at the corporate R&D trends in the United States today, it is concerning that industry is migrating away from research to concentrate more on development. Additionally, there is a growing concern that public corporations in the United States are reducing their focus on long-term discoveries in response to pressure from investors to focus on quarterly earnings.

This issue of short-termism in the United States has received a great deal of attention in the media in recent years. A National Bureau of Economic Research study titled "Killing the Golden Goose? The Decline of Science in Corporate R&D," published in early 2015, documents a shift away from scientific research by large corporations since 1980.[8] Researchers observed that the number of publications by corporately funded scientists have declined over the past several decades across a range of industries. The group analyzed changes in the level of research investment from 1980 to 2007 by more than one thousand firms engaged in R&D. They found that the total percentage of those firms publishing research papers fell from 17 percent to 6 percent during that time, whereas the percentage of patents filed increased from 15 percent to 25 percent. The researchers also found that the value attributable to scientific research has fallen, whereas the value attributable to technical knowledge (as measured by patents) has remained stable. These effects appear to be associated with globalization and a narrowing scope rather than changes in publication practices or a decline in the usefulness of science as a catalyst for innovation.

Corporations invest in R&D to foster innovation that leads to competitive advantage, increased market penetration, and profitability. Innovation is critical to business success. Studies show that business expenditure on R&D tends to be more closely linked to the creation of new products and techniques than R&D performed by government or academia.

It has become clear to company executives over time that successful innovation requires more tightly aligning R&D investment with business strategies. Researchers such as Gary Pisano and entrepreneurs such as Steve Blank are educating large companies on how the innovation process can help them operate internally with the speed, urgency, and success of early startups.[9]

Researchers at *Strategy + Business* publish a comprehensive review of corporate R&D trends each year. Over the years, they have identified core strategies that can improve a company's return on its R&D investment and witnessed some consensus regarding the factors that drive results. One of the key messages that has comes out of their analysis (including interviews with C-suite executives engaged in R&D) is that innovation leaders feel they have made real progress in more effectively leveraging their R&D investments, particularly by more tightly aligning their innovation and business strategies and by gaining better insights into customers' stated and unstated needs. According to the *Strategy + Business* 2014 global R&D survey, "44 percent of survey respondents noted that their companies are better innovators today than they were a decade ago, while another 32 percent said they are much better. Only 6 percent say they were doing worse."[10] The remaining 18 percent say they are neither worse nor better.

Recognition of the need for greater R&D spending on scientific technologies seems to have been growing in the U.S. corporate sector in recent years, and the latest R&D spending outlook by *Strategy + Business* found that, over the coming decade, businesses plan to shift their current R&D spending mix from incremental innovations to more new and breakthrough innovations (figure 2.6).[11]

Researchers at *Strategy + Business* note that, to capitalize on such a significant reallocation of spending, many companies will

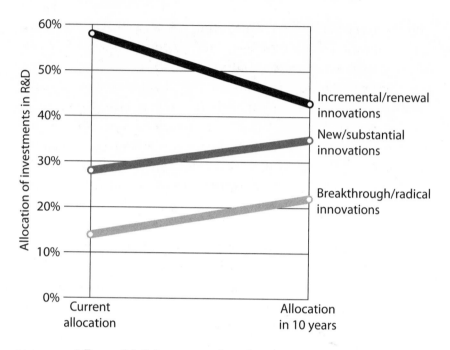

FIGURE 2.6 Future R&D investment plans. Survey respondents report expecting to shift their R&D investments mix from incremental to new and breakthrough innovations. *Source*: Barry Jaruzelski, Volker Staack, and Brad Goehle, "The Global Innovation 1000: Proven Paths to Innovation Success," *Strategy + Business*, October 28, 2014.

need to make major changes in their capabilities and approaches to innovation. This dovetails with the views of Pisano, Blank, and other intellectuals who work closely with corporations on innovation strategy.

Many corporations engaged in deep science are also active in the venture capital business. Established technology and biotechnology companies such as Intel, Google, Eli Lilly, Pfizer, among others, use their venture capital companies as a way to facilitate innovation. On paper, the case for establishing a corporate venture capital arm looks rather compelling—especially at a time like today when technological innovation is accelerating and the business landscape is shifting. Among the major advantages of a corporate venture capital fund is flexibility and faster-paced, more cost-effective progress than traditional R&D. These advantages can be highly beneficial to the innovation process if managed properly.[12]

While compelling on paper, the enterprise of corporate venturing is more likely to fail than to succeed, similar to financial venture investing. The median life span of corporate venture programs has historically been around one year. Even companies with successful venture capital funds have sometimes found it challenging to incorporate the knowledge gained from startup investments. Moreover, there has been a tendency for established companies to become interested when the venture capital business is booming and attempt to get into the business while the getting is good. This tendency only serves to exacerbate the booms and busts that occur in the venture capital business. Waves of corporate venture activity, like those that occurred in the late 1960s, the mid-1980s, and the late 1990s, correspond with booms in venture capital investments and venture-backed initial public offerings (IPOs).

Supporting this idea, currently, is the increase in corporate venture funds and the 79 percent increase in the number of companies making venture capital investments to 801 globally since 2011. Even as total venture capital investments doubled from 2011 to 2015, the amount invested by corporate venture funds quadrupled to $7.6 billion in the same period, according to the National Venture Capital Association.[13]

Deep Science: Progressing from Breakthrough to Commercialization

There are three phases in taking deep science from the research lab to market commercialization. Figure 2.7 depicts the value creation process of building deep science companies. In the early stages

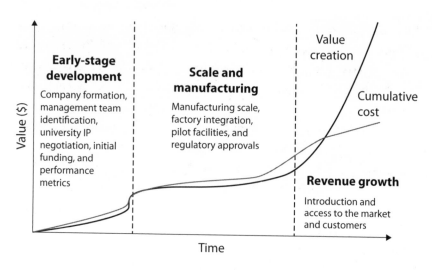

FIGURE 2.7 Deep science: from lab to market. *Source*: Harris & Harris Group.

of deep science investing, entrepreneurs develop the intellectual property and demonstrate that the technology can work in real-world conditions.

For most researchers in academic and federal research institutions, the goal of science is to publish papers. To publish, a successful observation or experiment must be documented and its reasons explained. If the output of that research is something tangible, often times, just a single example is created and described by the researcher, with the focus then turning to publishing the results. In venture capital circles, the initial discovery is often called the "hero device." It is the best observable output, created in the lab, operated under laboratory conditions, without a focus on cost or repeatability.

The next step is to take that initial discovery or output and prove first, that its creation can be repeated; second, that the performance witnessed in the lab can be repeated under real-world conditions; and third, that the product can be manufactured at a competitive marketplace price. It is in early-stage development that a company is typically formed, early capital invested, a management team formed, intellectual property secured, a lab established, and early metrics for performance defined and evaluated.

Often, there is a small value inflection point for proving the value of the technology at this stage. Most often, if there is value realized at this stage, it is an existing corporation that realizes the benefit of this new technology and pursues it as a new innovation or as a substitute for a technology it has previously brought to market. However, if the technology is more radical and the performance metrics substantially better than existing substitutes in the current market, the company may also solicit additional investment to develop the technology.

It is the next phase, scale and manufacturing, where the real bottleneck in deep science investing resides. Enabling deep science technologies usually requires large-scale manufacturing of some tangible object. In the life sciences, the output could be a therapeutic drug, a diagnostic tool, or a machine that sequences the human genome.

In each of these cases, there is a significant time lapse before a new technology is brought to real-world manufacturing and ultimately to scale. This is difficult. Multiple components in addition to the initial discovery are often needed to make a finished product. Software may be developed alongside a hardware innovation. Working on a repeatable process at an efficiency necessary to stimulate general adoption of the technology takes time. Successful scale and manufacturing takes a magnitude of resources greater than those required in early-stage development.

For example, in bringing the Tesla electric car to market, Elon Musk made multiple innovations in engineering and manufacturing automobiles, drivetrain innovations specific to electric vehicles, battery innovations (which are ongoing), and software developments to make the Tesla more like a computer on wheels than a traditional automobile. All these innovations then needed to be manufactured together and at sufficient scale to make Tesla successful in the marketplace. Elon Musk and Jeff Bezos's rocket innovations are even more extreme examples of such scale and manufacturing challenges, as well as the component innovations required to create a product that can begin to generate revenue.

Three historic areas of deep science investing have been semiconductors and electronics, energy, and the life sciences (health care and agriculture). The scale-and-manufacturing phase for electronics and semiconductors often takes five to seven years and can cost between $75 to $150 million before a new semiconductor product can be produced at the scale and price required by the market. Recent clean energy technologies, including solar power, wind power, and renewable chemicals and fuels, demonstrated a similar time line to the semiconductor industry, but with a cost of between $400 million and $1 billion to bring a new energy technology to the market at commercial scale. Therapeutic drugs dwarf both of these. Pharmaceutical firms spend approximately $1.8 billion to bring a new drug to market, including the failures along the way, while startup biotech companies have demonstrated they can bring a drug through the three phases of clinical trials for a minimum of $250 million.[14]

Scale and manufacturing are typically not the expertise of venture investors. Historically, large corporations developed this skill set as they matured and grew in size. Venture investors need the help of corporations during this phase. This phase proceeds best if the new technology can incorporate the manufacturing process used by an established industry. It is more difficult to engage corporate help if the process is entirely new or does not fit into their existing manufacturing processes. In the latter cases, the startup company usually must develop the manufacturing and scale itself.

During the scale-and-manufacturing phase of deep science development, value tends to increase less than cost. In most deep science innovation, except for drug development, value is difficult to realize during this phase, which translates into large outlays of capital without the corresponding increase in value for the capital being deployed. This makes it very difficult to attract additional investment in this period between early-stage development and revenue growth. In biotech investing, value inflection points can be realized earlier because of generally accepted value accretion points through the three phases of clinical trials. This has made biotech investing one of the few areas of deep science investing to perform well over the past five years.

In contrast to deep science investments, in software investments, a group of programmers can quickly develop code and then release the product to the market through the Internet. Interest can be ascertained by the number of users that begin working with the product or through the engagement of the market around the product. Then, investment dollars can be used to develop a revenue model and the company can begin selling to a large market that has been attracted to the product. A captive audience can often be determined prior to full launch.

For deep science technologies, however, that is not always the case. Demand for a product or technology usually occurs only once the product can be manufactured at scale. This means that larger investments are required prior to knowing if there is actually market interest for the technology. Additionally, market interest is often difficult to gauge ahead of widespread availability of the technology.

The final stage of bringing deep science from the research lab to market commercialization is revenue growth. It is during this phase of deep science investing that the value inflection point increases much more rapidly as revenue increases. For most deep science technologies, it is not until this third phase that value becomes recognized by many investors. For most deep science companies, significant amounts of capital must be invested prior to increasing revenue and first realizing profits. Before software began to "eat the world," this capital was available more generally, but capital seeking shorter-term time horizons has now migrated to those technologies that can create value inflection points over shorter periods of time and with less capital required ahead of revenue growth.

The one outlier in deep science investing is the drug development process, in which it can cost between $250 million and $1.8 billion to bring a new drug to market. This means that between $250 million and $1.8 billion is spent before a company knows if the drug will work and to what degree it will be adopted in the marketplace. Despite this cost, the biotech investment space has remained robust since 2000, albeit with temporary disruptions in the general market following the 2008 financial crisis.

There has remained a strong investing interest in new drugs because, if they are successful, there is usually a highly regulated and controlled, large, growing market. Additionally, regulation of the drug development process creates four very clear delineations that investors have used to gauge value: entry into phase I, phase II, and phase III clinical trials and then approval and first sales of the drug.

Prior to human (clinical) trials, a drug goes through preclinical trials. This phase equates to the early-stage development process illustrated in figure 2.7. During a new drug's preclinical development, the sponsor's primary goal is to determine whether the product is reasonably safe for human use and exhibits pharmacological activity justifying commercial development. The U.S. Federal Drug Administration's (FDA) role in the development of a new drug begins when the drug's sponsor (usually the manufacturer or potential marketer) tests its diagnostic or therapeutic potential in humans through clinical trials. At that point, the molecule's legal status changes under the Federal Food, Drug, and Cosmetic Act and officially becomes a new drug subject to specific requirements of the drug regulatory system.

After the FDA approves an investigational new drug (IND), the drug goes through three phases of development. In phase I, clinical trials are conducted to determine the drug's basic properties and safety profile in humans. Typically, the drug remains in this stage for one to two years. In phase II, efficacy trials begin as the drug is administered to volunteers of the target population. At the end of phase II, the manufacturer meets with FDA officials to discuss the development process, continued human testing, any concerns the FDA may have, and the protocols for phase III, which is usually the most extensive and most expensive part of drug development.

Because of the regulations surrounding the launch of a new drug entity, a viable marketplace has grown over time for investing in new drugs. Value inflection points have been realized as a new drug attains IND status and then at each of the three phases during the IND process.[15] Both private market investors and public market investors have recognized these value inflection points, which has led to the creation of a market for investors with different investors assuming risk at different stages of the drug's development.

Returning to our discussion of deep science investing, manufacturing and scale is the lynchpin between the early-stage development and revenue-growth phases, where value is often more generally realized. Gary Pisano and his Harvard colleague Willy Shih have provided some of the most vibrant thinking on this topic over the past decade. In their 2012 book, *Producing Prosperity*, Pisano and Shih describe the "modularity–maturity matrix" (figure 2.8).[16] Pisano and Shih discuss why the United States needs a manufacturing renaissance if it is to remain globally competitive. But this matrix also provides the answers for why manufacturing and scale are such important components of deep science innovation and commercialization.

The modularity–maturity matrix has four quadrants: process-embedded innovation, pure product innovation, process-driven innovation, and pure process innovation. The y-axis describes process maturity from low to high, and the x-axis describes the modularity of the manufacturing process from low to high. When modularity is low, the product design cannot be fully codified in written specifications, and design choices influence manufacturing. When manufacturing technologies are immature, the opportunities

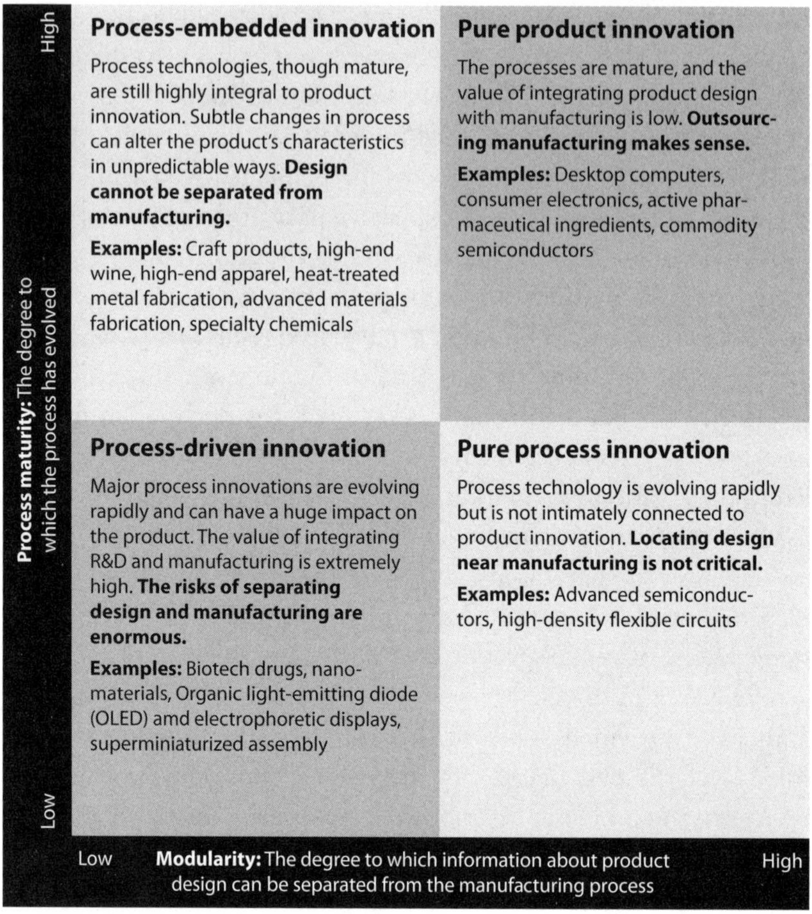

Process-embedded innovation

Process technologies, though mature, are still highly integral to product innovation. Subtle changes in process can alter the product's characteristics in unpredictable ways. **Design cannot be separated from manufacturing.**

Examples: Craft products, high-end wine, high-end apparel, heat-treated metal fabrication, advanced materials fabrication, specialty chemicals

Pure product innovation

The processes are mature, and the value of integrating product design with manufacturing is low. **Outsourcing manufacturing makes sense.**

Examples: Desktop computers, consumer electronics, active pharmaceutical ingredients, commodity semiconductors

Process-driven innovation

Major process innovations are evolving rapidly and can have a huge impact on the product. The value of integrating R&D and manufacturing is extremely high. **The risks of separating design and manufacturing are enormous.**

Examples: Biotech drugs, nanomaterials, Organic light-emitting diode (OLED) amd electrophoretic displays, superminiaturized assembly

Pure process innovation

Process technology is evolving rapidly but is not intimately connected to product innovation. **Locating design near manufacturing is not critical.**

Examples: Advanced semiconductors, high-density flexible circuits

High / Low — **Process maturity:** The degree to which the process has evolved

Low / High — **Modularity:** The degree to which information about product design can be separated from the manufacturing process

FIGURE 2.8 Pisano and Shih's modularity–maturity matrix. *Source*: Gary Pisano and Willy Shih, *Producing Prosperity: Why America Needs a Manufacturing Renaissance* (Boston: Harvard Business Review Press, 2012).

for improving the process are great and impact the final product and cost.

Most deep science innovations require a technology breakthrough before additional breakthroughs in how to produce, manufacture, or use the new technology. For this reason, deep science technologies are typically located in the lower left-hand quadrant of the modularity–maturity matrix. Major process innovations evolve rapidly as the technology is developed, and these complementary process innovations have a large impact on the product. Pisano and Shih cite examples like biotech drugs, nanomaterials, and electrophoretic displays. Occasionally, deep science technologies are developed where there is a higher degree of modularity, but most of the time, the process maturity is still low. Here manufacturing and scale can happen more rapidly because the process technology is not necessarily intimately connected with the product innovation. In such cases, the intimacy between process R&D and manufacturing is more important than the intimacy between product design and manufacturing; hence, locating manufacturing where there is the greatest opportunity for process innovation is sensible. Examples provided by Pisano and Shih include advanced semiconductors and high-density flexible circuits.

Most often, deep science investing requires that manufacturing and product design go together, but this translates into higher risk for investors as the product technology and the process technology are developed in parallel. This risk is the root cause for why scale and manufacturing are critical hurdles to bringing deep science breakthroughs to the market. A tremendous amount of time and money go into the scale-and-manufacturing problem presented by deep science innovation. Without the market's recognition of the value accretion in this stage, investing in deep science becomes very difficult in comparison to software business plans—even if the value ultimately recognized by breakthroughs in deep science dwarfs that of other innovations.

From Pisano and Shih's work, we may also infer that to have a vibrant innovation economy, there must be a diversity of ideas that move through the innovation channel from R&D, to early-stage investment, to scale-up, and then to commercial viability. If this innovation channel is narrow and diversity is lacking, the number of additional opportunities it creates will also necessarily be narrower. On the other hand, if the innovation channel is wide and diverse, it lays a scaffold for a wider set of innovations to develop. This is an argument posed by those who believe that it is important to keep some manufacturing in America. The innovation benefits that accompany manufacturing processes lead to new products and further process innovations.

America's early leadership in computers and video gaming was the direct result of the microprocessors revolution. By developing the infrastructure of the microprocessor and by understanding how an increase in processor speed could enable other follow-on innovations, American companies generally, and in Silicon Valley specifically, established leadership in the next generation of innovation that followed the microprocessor. The same may be said of the development of the personal computer.

The development of the underlying deep science innovations that enabled the Internet is another example of how a diverse array of technologies at a foundational level enabled a wealth of innovation in subsequent digital technologies. The underlying foundations for the Internet lie in multiple scientific fields, including math, physics, optical telecommunications, and information theory. This diversity of foundational knowledge became the substrate from which multiple additional industries developed, all of which aided the advancement of the Internet and software investments. Involving oneself in innovation at the ground floor often, although not always, leads to a leadership position in the ensuing industries that develop. That deep science innovation leadership has been important for the United States over the past century and certainly in the multiple advancements in the post–World War II economy.

Entrepreneurs and Venture Capital in the American Economy

Now that we have looked at the ecosystem that is the foundation of deep science innovation, we can turn our attention to the role of traditional venture capital in America. The investment activities associated with venture capitalists are vital for economic dynamism, and these activities facilitate an economic payoff from R&D that is vital to the innovation process.

The productivity of entrepreneurs and venture capitalists is, in fact, an essential feature of the capitalist economy that has fueled prosperity. As economist Joseph Schumpeter wrote in his 1942 book, *Capitalism, Socialism and Democracy*:

> But in capitalist reality as distinguished from its textbook picture, it is not that kind of competition which counts but the competition from the new commodity, the new technology, the new source of supply, the new type or organization—competition which commands a decisive cost or quality advantage and which strikes not at the margins of the profits and the output of the existing firms but at their foundations and their very lives.[17]

Schumpeter's book was published during a decade when electrical machines, appliances, automobiles, and new medical therapies were transforming the U.S. economy. Color television, the transistor, instant photography, the Turing machine, the Jeep, penicillin, Tupperware, disposable diapers, and duct tape were all invented in the 1940s.

Schumpeter's words gained more currency as the revolution in digital computing began. Additionally, venture capital began its expansion and transformation in the decades following Schumpeter's writings, culminating in the modern venture capital model prominent in Boston and Silicon Valley. Given the impressive track record of entrepreneurs and venture capitalists, there is a much greater appreciation among economists, executives, business leaders, and policymakers for the importance of technological innovation in the U.S. economy. Indeed, the term *disruptive innovation* (which has Schumpeterian overtones) has become part of modern-day business vernacular.

Entrepreneurs working in close collaboration with venture capitalists and corporations devoting resources to R&D have propelled economic dynamism in the United States over the past half-century. Entrepreneurs are agents of innovation, and venture capitalists facilitate that innovation by providing capital resources and experience. The entrepreneurial economy in the United States is based on converting ideas to inventions, and then into organized business activity. Entrepreneurs shift economic resources out of an area of low productivity into an area of higher productivity and greater yield, and venture capital is a key source of capital for those entrepreneurs.

Entrepreneurs funded by venture capital have proven to be a potent combination with respect to innovation and promoting economic dynamism in the United States. Indeed, a large share of the innovation witnessed in the second half of the twentieth century would not have occurred without entrepreneurs' initiative and the support of venture capital.

Innovation involves an enormous amount of uncertainty, human creativity, and chance. The traditional venture capital business model embraces these elements. Venture capital thrives on the risk-taking associated with uncertainty and human creativity. Some of the greatest American entrepreneurs of the post–World War II era benefitted from access to venture capital and close collaboration with venture capitalists.

Startup businesses launched by entrepreneurs that seek to commercialize scientific products and services often require a great deal of capital. Financial and corporate venture capitalists provide capital along with strategic guidance and management assistance. While the venture capital model is not perfect, it has proven to be a highly successful method for jumpstarting the innovation and commercialization of transformative technologies based in deep science.

Venture capital has played a significant role in financing some of the most dynamic and innovative companies in the world, including Amgen, Apple, Atari, D-Wave Systems, Genentech, Google, Intel, and Tesla. Venture financing has spawned dynamic industries, such as the microprocessor, video gaming, the computer, biotechnology, and the Internet.

Corporate venture investing also began to prosper in the 1970s and 1980s by first supporting some of the radical new technologies developed by deep science. However, over the past decade, as corporations' focus on their own bottom lines has intensified, R&D and deep science venture investing has given way to more of a focus on later-stage product development. As industries have become more institutionalized, the focus of venture capitalists has narrowed, and the support for deep science entrepreneurial ventures has declined.

Venture capital has had a profound impact on the U.S. economy. Although the domestic venture capital business is relatively young, venture-backed companies have risen to constitute one-fifth of the total value of all U.S. publicly traded companies. The comparatively brief history of venture capital in the United States shows quite clearly that this source of financing is a significant catalyst of economic dynamism.

Venture capital funding in the United States is a relatively small share of total capital employed by companies annually to sustain and grow businesses. Over the past fifty years, American venture capitalists have raised $600 billion and made thousands of investments from that pool of capital (figure 2.9). By contrast, over the same period, the private equity industry raised $2.4 trillion—four times as much. In 2014, the private equity industry raised $218 billion, almost ten times the $31 billion raised by the venture capital industry. Owing to these large numbers, it may be surprising to learn that venture capital funds invest in only 0.2 percent of new U.S. businesses. But do not let this statistic fool you.

The relatively small amount of venture capital invested annually belies its effect on the economy. Stanford University researchers Ilya A. Strebulaev and Will Gornall have documented that, over the

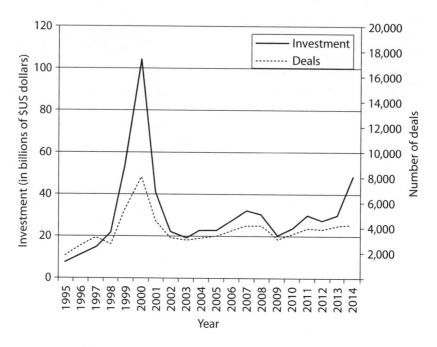

FIGURE 2.9 U.S. venture capital trends, 1995–2014. *Source*: Pricewaterhouse Coopers/National Venture Capital Association, Historical Trend Data, https://www.pwcmoneytree.com/HistoricTrends/CustomQueryHistoricTrend.

past thirty years, venture capital has become a dominant force in the financing of innovative American companies. From Google to Intel to FedEx, companies backed by venture capital have deeply transformed the American economy, as summarized in box 2.2.

The Economic Impact of Venture Capital

Strebulaev and Gornall compiled a database of over 4,000 public companies with a total market capitalization of $21.3 trillion. Venture capital–backed companies accounted for 18 percent, or 710 of the total, with a total market capitalization of $4.3 trillion, or 20 percent of the value of all publicly listed companies. Strebulaev

BOX 2.2

Venture Capital–Backed Companies as a Percentage of
Public U.S. Companies Founded Since 1979

  Total number: 43 percent
  Total market capitalization: 57 percent
  Total employees: 38 percent
  R&D: 82 percent

Source: Ilya A. Strebulaev and Will Gornall, "How Much Does Venture
Capital Drive the U.S. Economy?" *Insights by Stanford Business*, October
21, 2015, www.gsb.stanford.edu/insights/how-much-does-venture-capital-
drive-us-economy.

and Gornall observe that venture capital–backed companies tend
to be young and fast growing. While only accounting for a relatively small share of revenue (10 percent), venture capital–backed
firms provide 42 percent of total R&D funding.

That amount is more than a quarter of the total government,
academic, and private U.S. R&D spending of $454 billion. Ven-
ture capital–backed companies are also an important dynamic in
the labor market, employing four million people. The importance
of startups and venture capital–backed companies in the creation
of new jobs finds solid support from the U.S. Census Bureau's
business development statistics program. The data over the past
two decades show that virtually all new jobs created in the United
States came from startups.[18]

Strebulaev and Gornall point out that these statistics understate
the importance of venture capital because many of the large public
companies, like Ford, General Motors, DuPont, and General Electric, were founded before the creation of the venture capital business in the United States. To level the playing field, the researchers
redid their analysis using only those companies founded during or
after 1979 when the Prudent Man Rule was relaxed, bolstering

the venture capital business.[19] This method of analysis changed the results dramatically. Of the approximately 1,330 companies that were founded between 1979 and 2013, 574, or 43 percent, are backed by venture capital. These companies collectively compose 57 percent of the market capitalization and 38 percent of the employees of all such "new" public companies. Moreover, their R&D expenditure constitutes an overwhelming 82 percent of the total R&D of new public companies.

These results also illustrate the effect that changes in government regulation, such as the Prudent Man Rule, can have on the overall economy. It is also noteworthy that venture capital–backed companies make up a consistently high fraction of IPOs in the United States. According to Strebulaev and Gornall, over 2,600 venture capital–backed companies had their IPOs between 1979 and 2013. These made up 28 percent of the total number of U.S. IPOs during that period. The percentage of IPOs that were backed by venture capital varies by year. That percentage reached a high of 59 percent during the dot-com boom but has not been greater than 18 percent in each of the last twenty years.

Innovation and venture capital go hand in hand. In 2013, for example, venture capital–backed U.S. public companies spent $115 billion on R&D, up from essentially nothing in 1979. Today these venture capital–backed companies account for over 40 percent of the R&D spending by U.S. public companies. On average, a dollar of venture capital appears to be *three to four* times more potent in stimulating the patenting of intellectual property than a dollar of traditional corporate R&D.

The presence of a venture capitalist is also associated with a shorter time line for bringing an innovative product to market. Firms that are pursuing a strategy of innovation are much more likely to obtain venture capital, and to get it faster. In short, the impact of venture capital on innovation is significant.[20]

As Strebulaev and Gornall observe, the R&D spending associated with venture capital–backed firms produces value not just for those companies, but also for the entire country and the world through positive spillovers. Venture capital funding works to expand the economic pie and enrich not only the entrepreneurs and venture investors that back them, but provide benefits to customers and society at large that extend for many years and often over decades.

Strebulaev and Gornall's analysis supports the contention that a relatively small amount of venture capital creates a large number of dynamic companies, which, in turn, fuel innovation and job creation over time. Venture capital has been an important catalyst of economic dynamism in the United States precisely because venture capitalists specialize in investing in innovative companies with major potential for growth. By employing a relatively small amount of financing, venture capitalists foster the development and penetration of revolutionary and transformative innovations in the economy.

In chapter 3, we will turn our attention to the history of American venture capital and its historical role in bringing deep science technologies to the market. Through this process, we will gain insight into how venture capital works and its role in helping to create a payoff to R&D.

3

Deep Science and the Evolution of American Venture Capital

FINANCIAL VENTURE INVESTING HAS been in use since the beginning of organized commerce. The Dutch, Portuguese, Spanish, and Italian governments that sponsored transatlantic voyages and trade during the sixteenth century were taking part in a form of venture investing. Investors that backed the technologies emerging from the first and second Industrial Revolutions in England and the United States were also participating in a form of venture investing. However, venture investing as an organized, professional institution was not conceived until the Great Depression of the 1930s and not "born" until the second half of the twentieth century.

One of the best compiled histories detailing the formation of American venture capital and of American Research and Development (ARD), the first American venture capital firm, has been written by Spencer Ante in his book *Creative Capital*. Much of the early part of this chapter draws from his work.

A result of American Depression-era tax policies was the lack of risk in the American economy of the 1930s and 1940s, and thus an economy that lacked innovation. A series of Revenue Acts between 1932 and 1937 restricted small companies from building up their capital from earnings. These Revenue Acts limited wealthy individuals from investing in small companies. The result was that funds flowed disproportionately into conservative investment trusts, insurance companies, and pensions.

As Spencer Ante describes, this situation led to New York University finance professor Marcus Nadler commenting at the 1938 Investment Bankers Association Conference, "If investors throughout the land, large and small, refrain from purchasing unseasoned securities of a young industry and refuse to take a business man's risk, where will new industries obtain needed capital, and would not such a development slow down the economic progress of the country?"[1]

In the 1930s, New England–based industrial economists began brainstorming how to address the problem of a regulated economy that favored less risky investments. Of the groups that organized, one of the most prominent was the New England Council. This group understood that New England's universities and industrial research centers, including MIT, were valuable assets that could be used more effectively.

Around the same time, Karl Taylor Compton, a Princeton physicist, became president of MIT. In 1934, he proposed a program titled "Put Science to Work." This program focused on developing new industries founded in scientific innovation. The New England Council supported Compton's ideas and in 1939 formed a "new products" committee to examine how new products might help reverse the decline of certain New England industries.

This new committee brought together a few very progressive minds, including Compton, Georges Doriot, Ralph Flanders, and Merrill Griswold. Georges Doriot was placed in charge of a subcommittee called "Development Procedures and Venture Capital."[2] The conclusion reached by this subcommittee was that there was capital available for new ventures, but that an organized approach to obtaining qualified technical analysis for these new ventures was necessary to provide the diligence necessary to attract such capital.

World War II turned out to be a great stimulus for entrepreneurism in the United States. War encouraged risk taking in the areas of new technologies and new methods of production. Spencer Ante details three such ventures in *Creative Capital*. First, the rise and success of the synthetic rubber industry resulted from the acceleration of what had been an unproven technology prior to the war. Second, the success of technologies developed for the war encouraged many investors to pursue a greater level of risk because the potential for return was now more palpable. Third, the Allied victory removed the last traces of Depression-era timidity, and in its place a newfound self-confidence and experimentation began to grow.[3]

World War II also changed the role the federal government played in basic research, which is the systematic study directed toward greater knowledge or understanding without specific applications or commercial products.

Prior to the war, U.S. basic science expenditures were less than $40 million annually. By 1943, however, U.S. basic science expenditures at research universities and foundations had more than doubled to $90 million a year in contracts.[4] In the years following the war, the government spent 10 to 15 percent of its annual R&D budget on basic research, about half of which was spent in universities where basic research accounted for approximately two-thirds of all research.[5]

The rise of the research university after World War II is largely the product of Vannevar Bush's postwar vision, described in a 1945 report to the president titled *Science: The Endless Frontier*. Bush was an American engineer, inventor, and science administrator who headed the U.S. Office of Scientific Research and Development (OSRD) during World War II. The OSRD founded the Manhattan Project. Bush defined the federal government's support for R&D as the "postwar social contract." The central premise of this "contract" was that the federal government would support university research, permitting universities a high degree of self-governance and intellectual autonomy, in return for which their benefits would be widely diffused through society and the economy.[6]

With the dissolution of the OSRD after the war, Bush and others had hoped that an equivalent peacetime government R&D agency would replace the OSRD. Bush felt that basic research was important to national survival for both military and commercial reasons; technical superiority could be a deterrent to future enemy aggression. In *Science: The Endless Frontier*, Bush maintained that basic research was "the pacemaker of technological progress. New products and new processes do not appear full-grown. They are founded on new principles and new conceptions, which in turn are painstakingly developed by research in the purest realms of science!"[7]

Modern organized venture investing can trace its origin to immediately after World War II when MIT president Karl Taylor Compton publicized his plan to create a new type of financial firm that would finance the development of technical and engineering companies. Compton approached the New England Council's new products committee to convince them to start a new venture capital firm. The result was the formation and incorporation of the American Research and Development Corporation on June 6, 1946, by Ralph Flanders, Frederick Blacksall Jr., MIT treasurer Horace Ford, and Georges Doriot, who would be chairman of the board of directors.

ARD: A Case Study for Institutional Venture Capital

During the first half of 1946, two other venture investing organizations also formed. The East Coast Whitney and Rockefeller families formed J. H. Whitney and Company and Rockefeller

Brothers Company, respectively. However, ARD was the first to pursue capital from nonfamily funds, primarily institutional investors such as insurance companies, educational organizations, and investment trusts. As the first public venture capital firm, ARD also sought to democratize entrepreneurship by focusing on technical ventures and by providing the intellectual leadership for this small, nascent community.[8]

Venture investing began with ARD in 1946 and its initial capital under management of $3.5 million. Of this sum, just over $1.8 million was provided by nine financial institutions, two insurance companies, and four university endowments: MIT, Rice University, the University of Pennsylvania, and the University of Rochester. Individual stockholders contributed the rest of the capital.[9]

The idea of venture capital was so new that ARD's founders were forced to re-engineer financial regulatory structures in order to make the idea viable. ARD had to seek multiple exemptions under the Investment Company Act of 1940 from the U.S. Securities and Exchange Commission (SEC) before it could even offer its stock to the market.

For instance, Congress and the SEC prevented investment companies from extending their control through investment pyramids, as was done frequently in the 1920s. Consequently, one section of the Investment Company Act stated that an investment company could not own more than 3 percent of another investment company's voting stock. In the case of ARD, this would have prevented the Massachusetts Investors Trust from buying a large block of ARD stock. This rule is still in effect today, keeping any single investment fund from owning more than 3 percent of a business development company and any single investment company (even those with multiple funds) from owning more than 10 percent of an investment company.

Georges Doriot was born in France in 1899 and immigrated to the United States in the 1920s to earn an MBA, following which he became a professor at Harvard Business School. He became a U.S. citizen in 1940 and the following year was commissioned a lieutenant colonel in the U.S. Army Quartermaster Corps. As director of the military planning division for the quartermaster general, he worked on military research, development, and planning and was eventually promoted to brigadier general.

Doriot set the tone for the type of investment ARD would be involved in early on: "ARD does not invest in the ordinary sense. Rather, it creates by taking calculated risks in selected companies in whose growth it believes."[10] ARD's philosophy was to provide more than just guidance for monetary investment. It also provided managerial assistance and technical advice when necessary. ARD's initial six investments over its first year and a half demonstrate the type of far-ranging deep science it would fund:

Circo Products: a Cleveland-based company developing a method to melt car engine grease by injecting a vaporized solvent into automobile transmissions.

High Voltage Engineering Corporation: an MIT-founded company developing a two-million-volt generator eight times more powerful than existing X-ray machines.

Tracerlab Incorporated: a commercial atomic-age company, founded by MIT scientists, selling radioactive isotopes and manufacturing radiation detection machines.

Baird Associates: a Cambridge, Massachusetts, company making instruments used in the chemical analysis of metals and gases.

Jet-Heet Incorporated: a New Jersey–based company developing a household furnace based on jet engine technology.

Snyder Chemical Corporation: a company developing new resins for the paper and plywood industries.

In 1951, ARD reiterated its philosophy of taking calculated risks by creating new companies. This time, it used as an example its 1951 investment in Ionics, an MIT-founded company that demonstrated a new membrane that desalinated seawater more cheaply than any other existing technology and was first adopted in Coalinga, California, to replace water Coalinga had previously shipped in by railway.

In 1952, the growing success of High Voltage Engineering and Ionics convinced ARD's directors that the best opportunities were in early-stage investments in technology companies. These were its riskiest investments but the ones with the potential to generate the greatest financial returns.

In 1957, the U.S. government would come to recognize the importance of early investing in technology and the work ARD was doing after the Soviet Union surprised the world by successfully launching the Sputnik 1 satellite. Prior to this Soviet achievement, the United States had believed it was the world's leader in missile development and space technology. Like World War II, the Sputnik launch marked a profound turning point in the history of American innovation.

Over the next year, the U.S. government implemented a number of federally sponsored programs, such as the Department of Defense's Advanced Research Projects Agency of 1958, which forged a high-technology, entrepreneurial economy. That same year, President Dwight Eisenhower signed the bill that created the National Aeronautics and Space Administration (NASA), and, as a result, Congress dramatically increased funding for scientific research.

Sputnik also galvanized public support for venture capital, and Eisenhower signed the Small Business Investment Act of 1958, appropriating $250 million to start the Small Business Investment Company (SBIC) program, which offered tax breaks and subsidized loans to entrepreneurs.

ARD came of age in the late 1950s and 1960s. From the late 1950s through 1962, the United States experienced its first boom in high-technology stocks. Sherman Fairchild, the founder of Fairchild Camera and Instrument, was profiled in a 1960 *Time* magazine cover story that highlighted him as the "epitome of the new scientist–businessman–inventor."[11] ARD's portfolio was dominated by new issues trading over the counter, as ARD had been using the public markets to help fund its companies since its founding. Also in 1960, ARD invested in Teradyne, a company founded by MIT classmates Alex d'Arbeloff and Nick DeWolf. Teradyne created "industrial-grade" electronic test equipment that would become key to the growing semiconductor industry.

It was also in the 1960s that ARD's first "home-run" investment, Digital Equipment Corporation (DEC), began to blossom. DEC is generally recognized as the company that brought the first successful minicomputer to the marketplace in the 1960s. ARD's initial $70,000 investment in DEC had grown in value by over five hundred times after the company's initial public offering in 1968. DEC proved that venture capitalists could generate enormous wealth by backing the leader of a hot, new, innovative business.

ARD exemplified the venture industry's foundation over the coming decades in three key ways. First, ARD demonstrated that the riskiest investments could prove the most rewarding and that the greatest capital gains could be earned in the youngest companies. Second, it showed that most venture investing is not built upon overnight success, but on the steady growth of soundly based, well-managed companies. Third, ARD proved that capital investment in deep science technology was rewarding, because in these specialized technical areas, products are patent protected, thus making it easier for small companies to compete with large organizations.

The 1960s, however, also brought light to the first signs of ARD's ultimate downfall: its structural problems. As a public investment company, ARD was continuously lacking understanding and sympathy from the SEC. Throughout its existence, it fought with the SEC over investor ownership guidelines, compensation, valuation, employee ownership in its investee companies, and the inclusion of proprietary financial information from its subsidiaries in its own financials, making it less likely that small startup companies would want investment from ARD.

Although ARD's greatest difficulties as a venture capital firm were structural, its disintegration and the industry's realignment toward a private partnership structure can be traced to the departure of William Elfers, an ARD employee and Doriot's potential successor, in 1965. Elfers joined ARD in 1947 and, within a year, had been assigned the responsibilities of director, treasurer, and manager of ARD's investment in Flexible Tubing.[12] By 1951, Elfers had turned Flexible Tubing around, and he would go on to similarly save many other troubled investments over the years.

Elfers left ARD to form his own venture capital partnership, Greylock Capital, believing that the private partnership model solved many of the issues that the regulatory structure of ARD continued to face. Greylock was one of the first private venture firms to raise capital from several families, rather than a single limited partner like J. H. Whitney and Company and the Rockefellers had done. In 1965, Elfers raised $5 million from five wealthy families, including the Watsons of International Business Machines (IBM), Warren Corning of Corning Glass Works, and Sherman Fairchild, founder of Fairchild Semiconductor.

Greylock was not the first limited partnership for venture capital, though. In 1959, the first limited partnership was organized in Palo Alto, California, when William ("Bill") H. Draper III, a student of Doriot's at Harvard, formed Draper, Gaither and Anderson. The firm was re-formed in 1962 as the Draper and Johnson Investment Company with $150,000 from the Draper and Johnson families and $300,000 of SBIC money. Additionally, in 1961, Arthur Rock, an investment banker and student of Doriot's at Harvard, and Thomas ("Tommy") J. Davis Jr., a real estate investor, formed venture capital firm Davis and Rock, the industry's second significant partnership.

But in 1972, Greylock made a decision that would forever change the nature of the venture capital industry: to grow by forming a new partnership, rather than by raising capital into its existing fund. This permitted Greylock to increase the ownership of the younger general partners in the new fund. It also made it easier to admit new limited partners, because there would be no issue of valuing the existing portfolio companies for new investors. Finally, providing a fixed start date and end date made it easier to report performance. This ten-year fund lifetime has become a fixture in the venture community.

U.S. Deep Science Venture Capital Expands Westward

In the 1960s, the East Coast continued to control the venture industry with the largest pools of capital thanks to ARD, Greylock, the Rockefeller Brothers Company, and Fidelity Ventures, but the tide was beginning to turn, especially in the wake of World War II.

Prior to World War II, the northeastern United States enjoyed a regional advantage owing to its university leadership and financial superiority. MIT was the country's top science and engineering university and Harvard the top business school. New York was the financial capital of the world. Even after World War II, the northeast retained its technological advantage. Two important government research labs, the Radiation Laboratory and the Lincoln Laboratory, were run out of MIT. ARD-backed DEC traces its founding to the Lincoln Laboratory.

But during the 1960s, the West Coast began to take over the deep science technology industry. Key factors were the California Institute of Technology's growing reputation as a leader in cutting-edge science and technology and the leadership of Stanford University professor Frederick Terman. Terman studied engineering under Vannevar Bush at MIT, but took his first faculty position at Stanford in 1926. In the ensuing years, he grew frustrated seeing his top students flee to the East Coast for employment. For example, in 1934, two of his top students, David ("Dave") Packard and William Hewlett, went east after graduation, with Packard enrolling in management training at General Electric and Hewlett beginning his graduate work at MIT. However, Terman encouraged them back to Stanford with fellowships and part-time jobs.

One thing Terman enjoyed was taking recruits on tours of local electronic firms, which provided the entrepreneurial boost to recruit the top minds that Terman sought. And it was Terman who encouraged Hewlett and Packard to set up the Hewlett–Packard Company (HP) in 1939, working out of the garage in Dave Packard's back yard.

At the behest of Vannevar Bush, Terman worked at Harvard's Radio Research Laboratory during World War II, but, following the war, he returned to Stanford as dean of the School of Engineering. In this post, he continued to mentor a generation of leading researchers in their move into the private sector. This culminated in 1956 when Terman helped Nobel Prize–winner William Shockley recruit some of the best minds for the Shockley Semiconductor Laboratory, a division of Beckman Instruments.

Much of the rise of venture capital on the West Coast can be traced back to either the leadership of Doriot at ARD or to the time when the "traitorous eight" departed Shockley Semiconductor for Fairchild in 1957. These eight men were Julius Blank, Victor Grinich, Jean Hoerni, Eugene Kleiner, Jay Last, Gordon Moore, Robert Noyce, and Sheldon Roberts. Arthur Rock, at the time a New York investment banker with Hayden, Stone and Company, learned of the traitorous eight through Eugene Kleiner, whose father had an account with Hayden, Stone and Company. Rock was unsuccessful finding any company to back the group until he convinced inventor Sherman Fairchild to invest in them, thus founding Fairchild Semiconductor in September 1957.

Eugene Kleiner later went on to co-found Kleiner Perkins (now called Kleiner, Perkins, Caufield and Byers) in the early 1970s with Tom Perkins, an HP alumnus and friend of Doriot who had previously been interviewed to succeed Doriot at ARD. Kleiner, Perkins, Caufield and Byers is well known for investing in Tandem

Computers and Genentech. Donald ("Don") T. Valentine, who led sales and marketing at Fairchild, went on to found Sequoia Capital in 1972, which is now famous for financing Atari and Apple. Arthur Rock, who had partnered with Tommy Davis in 1961 to establish Davis and Rock, put together the financing deal for Intel in 1968, as Gordon Moore and Robert Noyce decided to depart Fairchild. Rock raised the first $2.5 million and wrote the business plan for Intel.

Other West Coast venture firms that began to emerge in the late 1960s and early 1970s include the Mayfield Fund, founded by Tommy Davis after splitting with Arthur Rock; Sutter Hill Ventures, founded by Bill Draper and Paul Wythes; and Institutional Venture Associates, founded by Reid Dennis.

The evolution of early venture capital in the United States is depicted in figure 3.1.

In its earliest days, the focus of venture capital was on bringing deep science technologies to the market, as evident from the early investments of ARD, Greylock, Davis and Rock, Kleiner Perkins, and Sequoia.

As we will see in the next chapter, however, venture capital has increasingly migrated away from the deep science business model toward software. This migration has major economic implications, given that deep science had played an integral role in fostering innovation and prosperity in the past. Before turning to the changing nature of venture investing and to the recent diversity breakdown in venture investing, though, we will spend some time examining a framework for venture investing and describing some of its properties.

Types of Venture Investing

Following in the footsteps of Schumpeter and Christensen, Harvard Business School professor Gary Pisano has created a framework

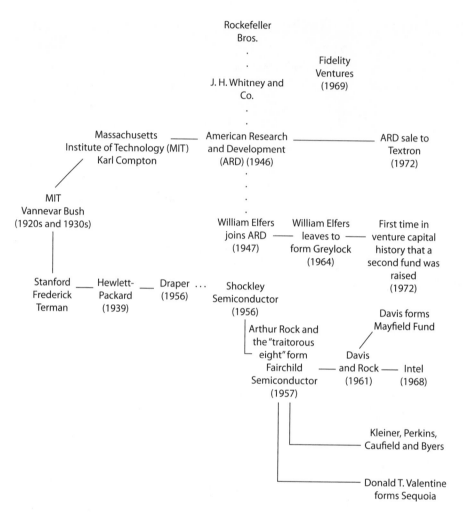

FIGURE 3.1 The evolution of early venture capital in the United States.

that is useful for assessing the types of venture investing that foster innovation and, in turn, economic dynamism (figure 3.2). Pisano's framework places new venture investments in one of four quadrants: (1) disruptive, which requires a new business model but leverages existing technical competencies; (2) architectural, which requires both a new business model and new technical

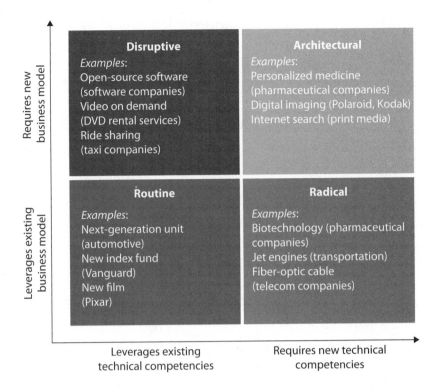

FIGURE 3.2 Pisano's framework for defining venture investing. *Source*: Adapted from Gary P. Pisano, "You Need an Innovation Strategy," *Harvard Business Review*, June 2015.

competencies; (3) routine, which leverages both an existing business model and existing technical competencies; or (4) radical, which leverages an existing business model but requires new technical competencies.

Venture capital has been most successful generating outsized returns when it invests relatively early in companies that are operating in the disruptive or architectural quadrant. Rather than trying to penetrate mature markets, in the disruptive or architectural quadrant, venture capital firms become involved relatively early, although not always first, in the formation of these new markets or industries. Biotechnology is a successful example of venture capital that fits into a lower quadrant, the radical quadrant. Its success can be attributed to how it vastly increased the mechanisms, efficacy, and targeting for therapeutic intervention despite being bound by a rigid regulatory framework dictating how it could enter the market.

The Pisano framework sheds light on recent venture capital failures, including the wave of investment in clean energy and nanotechnology during the first decade of the twenty-first century. Clean energy technologies are a substitute technology in a mature energy industry where there is a lot of competition and other difficulties associated with penetrating the market. Nanotechnology is a general-purpose technology that has great potential to foster innovation. That said, venture capital backing of nanotechnology has struggled because new markets and industries were not created, nor were the markets changed dramatically by nanotechnology products. Nanotechnology is neither a market or industry at all, nor has it created any new industries as of the time of writing, although the emergence of 3D printing, quantum computing, and new electronic materials have the potential to significantly change industries in the future. Looking forward, the breakthrough of inexpensive genetic sequencing could be a technology that enables a transformation in how health care is practiced, possibly leading to the realization of the promises of precision medicine.

The Pisano framework also sheds light on some of the early venture capital successes that required new business models. Clearly, many of the famous figures who started the West Coast venture industry trace their start to an understanding of how the microprocessor was going to establish and revolutionize an industry. But firms like Sequoia and Kleiner Perkins also realized that new forms of entertainment and new computing industries would be enabled by this microprocessor revolution.

Hence, in 1976, when two engineers from the early Silicon Valley engineering firm Ampex needed to grow Atari, they turned to Valentine. Valentine organized a round of capital that included the Mayfield Fund and Fidelity Venture Associates, now run by Henry Hoagland (who had left ARD). Valentine realized that these new powerful silicon chips could enable the type of computing power that could transform gaming from an era of pinball machines and penny arcades into an electronic realm. Atari had invented and marketed the game "Pong" prior to any investment, but the syndicate led by Valentine permitted the development of games such as "Breakout," "LeMans," and "Night Driver." The electronic gaming industry today continues to benefit from the increasing computational power that allowed Atari's takeoff.

Thomas ("Tom") Perkins and his colleague James ("Jimmy") Treybig both understood that growth in the power of microprocessors was also going to empower the computer industry. Backed by Kleiner Perkins in 1974, Perkins and Treybig founded Tandem Computers. The company was going to manufacture fault-tolerant computers that would continue to operate at a reduced level even when part of the system failed. The belief was that these machines could be marketed to banks and other financial firms to power early stock trading systems and automated teller machines (ATMs). By 1980, Tandem was ranked as the fastest-growing public company in America, and its partnership with Citibank drove its revenue growth.[13]

Don Valentine also realized that the growth in the power of microprocessors was going to create a whole new industry in personal computing. In 1977, Steve Jobs, designer of "Breakout" while at Atari, approached Bushnell to invest in a minicomputer company he and Steve Wozniak were launching. Bushnell referred Jobs to Valentine, who made introductions to A. C. ("Mike") Markkula Jr., who invested $250,000 of his own money, and the three formed Apple Computer. In 1978, the first venture capital dollars were invested by Sequoia, Venrock (the venture arm of the Rockefeller family), and Arthur Rock.

In 1974, Kleiner, Perkins, Caufield and Byers hired Robert Swanson, a twenty-seven-year-old MIT graduate with a bachelor of science in chemistry and an MBA from the Sloan School of Management. Swanson became interested in the emerging field of biotechnology, particularly the ability to manipulate the genes of microorganisms. Through Kleiner Perkins, Swanson had become involved with a company called Cetus, which had automated screening for microorganisms, but Cetus was unwilling to tackle the more ambitious task of splicing genes.

This led Swanson to Herbert Boyer, a forty-year-old biochemistry and biophysics professor at the University of California, San Francisco. Boyer had co-developed a technique to engineer drugs by splicing DNA from one organism into the genes of another. The idea that microorganisms could be used to make genetically modified products led to the establishment of Genentech in 1976. After convincing Kleiner Perkins that the risk could be reduced by subcontracting the initial research to University of California, San Francisco; the City of Hope medical research center; and the California Institute of Technology, Genentech raised its first venture investment from Kleiner Perkins.[14] The biotechnology industry went mainstream four years later, when Genentech went public in 1980.

The Power Law Distribution of Venture Investing

Venture funds invest in companies that are statistically referred to as "tail distributions." Venture funds realize that they may lose money on investments many more times than they make money. Additionally, investors do not invest all the money at once, and thus, over time, they can begin to recognize the lemons ripening ahead of the plums.

This awareness has led to what has become defined as the "power laws" of venture investing. As Peter Thiel writes, "The biggest secret in venture capital is that the best investment in a successful fund equals or outperforms the entire rest of the fund combined."[15]

To be successful, a venture firm has to nurture many disruptive ideas. Yet, only a small subset, over a span of five to fifteen years, will find the right market at the right time and execute heroically to become companies that transform their respective markets. These "home runs" are the companies that generate the majority of all returns in a venture capital portfolio.

The mathematical and economic language used to describe the behavior of venture returns versus those of standard operating companies is that venture operates on a power law distribution, not a normal distribution. This means that a small handful of investments—or even a single investment—can radically outperform all the others. This outperformance generates growth that may take time to realize but that can cause rapid growth in the fund when it does occur.

Owing to the power law nature of venture capital investing, and because a single investment can outperform the entire rest of the fund, the appropriate way to deploy capital in a standard ten-year venture fund is to make sure each investment made has the potential to return the fund. Thus, if the fund is $200 million in size, and that fund plans to reserve $20 million for ten investments over the ten years, then each $20 million deployed should be invested in a company that can return $200 million. This equates to each investment having the potential to be a ten-times return.

Because venture investors often syndicate deals, meaning they have other investing partners, this also means that if that $20 million ultimately ends up owning 30 percent of the company, the company in which that $20 million is deployed must ultimately be valued or sold for greater than $670 million. This is why venture capital investing is a "home-run" game. To put this into perspective, for a sample of over seven thousand company sales transactions between 1970 and 2006, only 9.7 percent of all merger-and-acquisition (M&A) sales were valued in excess of $100 million.[16]

The Changing Nature of U.S. Venture Capital (2000–2015)

Venture capital began to shift dramatically around the turn of the twenty-first century (figure 3.3). Tremendous inflows of capital into venture capital as an asset class further institutionalized the industry prior to the dotcom collapse and the collapse of the economy in 2000, permitting many firms to rapidly expand. With greater access to capital, venture capital firms began to change, and the number of roles within a venture capital firm began to expand dramatically. A new legion of employees, paid for by the greater fees that came from managing more capital, began to fill out the ranks of many of the premier venture firms. Many of these new roles were populated by individuals with both technical degrees and business school experience, but with little experience in building startup companies.

As these larger funds recovered from the dotcom crash, many turned their attention back to deep science companies, investing in two popular trends in the early 2000s: nanotechnology and clean energy. Building these types of company was a return to the pre-dotcom, pre-software days of deep science. However, there was an important change. The venture capital funds now had institutionalized staffs that often had neither the skills to understand the science behind these companies nor the time frames or operational experience needed to build these companies. The firms that remained focused on software and digital technologies enabled by the Internet after the dotcom crash were more successful.

Additionally, with fresh, increased funds, the inflow of capital was too much to manage for managing general partners, so associates, senior associates, and junior partners were now making and managing new investments, but often without the authority to vote for the fund partnership on the investment or on the future capital needs of these companies. This situation created a

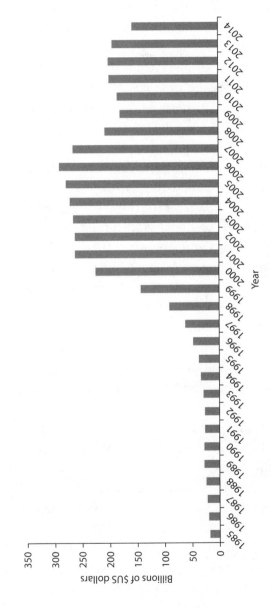

FIGURE 3.3 Capital under management in U.S. venture funds, 1985–2014. *Source:* National Venture Capital Association.

reporting system that was harder to monitor than the more nimble system that had existed previously. It also meant that the person interacting with the company might not be the person with a controlling vote on future funding or the ultimate decisions for that company. The joke within startups became that if the venture capital representative at the meeting did not have the power to vote or persuade his or her partnership, then he or she was there only for a free lunch.

The tremendous inflow of capital also meant that managing a successful venture capital fund became an exercise in raising ever-larger funds every two to three years. Although a typical fund survives for ten years, before capital needs to be returned to the limited partners of that fund, the initial investments are often made in the first few years. After these initial investments are made, a successful general partnership can then raise another fund and receive fees both for the initial fund, as it waits for these initial investments to mature, as well as a second fund, in which it is making new initial investments.

This permits the partners to oversee multiple funds at the same time, thus receiving an increased cumulative management fee. This process is the culmination of the innovation for which Elfers left ARD to form Greylock. Venture capital thus became a business of acquiring ever-greater assets. Figure 3.4 shows how venture capital funds grew in size beginning in the late 1990s through the early 2000s.

The hangover from the dotcom crash, heightened by the 1990s changes discussed, led to one of the worst-performing decades for venture capital between 2001 and 2010. This is evident in a comparison between different asset classes and their returns for this period in relation to previous periods (table 3.1). Venture capital, one of the riskiest asset classes, actually performed worse than most other asset classes over the course of this decade.

However, the venture industry is an intelligent (and perhaps overly optimistic) industry, which quickly moved to make changes.

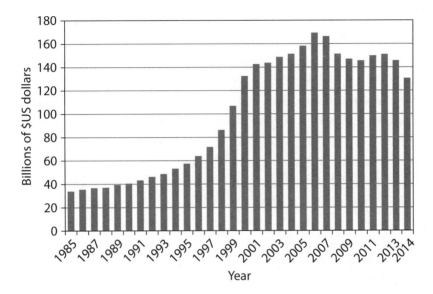

FIGURE 3.4 Growth in size of U.S. venture capital funds, 1985–2014. *Source*: National Venture Capital Association.

In order to deploy capital and raise an additional fund within three years, a fund has to seek companies that can progress rapidly within that time frame and that can return capital earlier than many of the deep science companies the venture industry was founded upon.

Table 3.1 Comparison of Asset Class Returns: Asset Class Returns for Period Ending December 31, 2010

	10 Years	15 Years	20 Years
U.S. Venture Capital Index*	–2.0	34.8	26.3
By stage			
Early stage	–3.3	46.1	25.6
Late and expansion	1.7	15.7	23.5
Dow Jones Industrial Average Index	3.2	7.9	10.3
NASDAQ Composite Index	0.7	6.4	10.3
S&P 500 Index	1.4	6.8	9.1

*Cambridge Associates U.S. Venture Capital Index
Source: Cambridge Associates.

Lucky for the venture industry, around the same time that funds were flowing into venture capital as an asset class and funds were growing and becoming institutionalized, a new major industry created by venture capital was emerging: digital technologies enabled by Internet software. This new industry fit the evolving venture capital model perfectly. Small initial investments were required to write software code that could create a product that would attract a large base of potential users. Users could be monitored easily, and those software applications that attracted large user bases could then be used to drive a revenue stream for that large user base. Large amounts of venture capital were required for growth, but only after investors knew there were tremendously large user bases that heretofore could not have been organized as widely or as quickly as the Internet now allowed.

We contrast this with technological progress in segments such as electronics, semiconductors, computing, life sciences, and energy, where a much larger initial investment in the technology is typically made long before a user base can be established and long before the product can be commercialized in an established market. As discussed previously, it is not uncommon for a biotechnology firm or a semiconductor firm to take five to ten years and require more than $100 million of invested capital before a product reaches the market. It can often be ten to twenty years before rapid adoption occurs for new technologies in the deep sciences. This reality does not work well for demonstrating progress in three years and raising the next fund. In the short term, this situation also makes it difficult to compete against investments built from software innovation.

Thus, the large amount of capital that entered the venture capital industry beginning in the late 1990s and early 2000s migrated very rapidly to software investments. Of the firms that did not make this migration, approximately two-thirds went out of business in what became one of the most difficult venture environments in history between 2001 and 2010. The venture industry after 2011 looks very different from the venture industry of the 1990s and earlier.

The Middle Class Squeeze

In sync with the diversity breakdown over the last decade, there has been a significant reduction in venture funds that have the capital to build these types of deep science companies. We have seen a large decrease in the $100 million to $500 million fund size, the historical size for building life sciences and physical sciences companies. This fund size has now decreased to 18 percent of all funds, down from 44 percent in 2008 (figure 3.5).

Deep science investing can be done effectively from a mega-fund. However, it is difficult to do early-stage investing from a mega-fund, as the time and effort required to make these initial, small investments do not justify the return these initial investments can make to a fund of that size. Therefore, most mega-funds must focus primarily on late-stage investments. This then raises the question of who will perform the early-stage work. Historically, the $100 million to $500 million funds invested from the early stage to the late stage in the deep sciences. With these funds

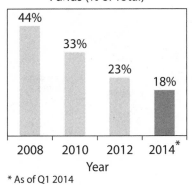

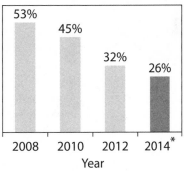

FIGURE 3.5 The shrinking venture capital middle class. *Source:* Q2 2014 Pitch-Book U.S. Venture Industry Date Sheet, Atelier Advisors.

decreasing from 44 percent to 18 percent, there are very few funds left to focus on early-stage deep science investing.

But as figure 3.6 shows, the dearth of $100 million to $500 million funds has been offset by a large increase in angel funds. The rise in angel investing is a good sign for early-stage investing in new innovation. However, the average angel fund investment (with multiple angels banding together) is approximately $650,000. Sixty percent of all deals and 55 percent of all investment dollars go into the web, mobile, telecommunications, and software sectors; 19 percent of deals go into health care; and the remaining 26 percent go into other segments, including advanced materials, semiconductors, and business services.[17]

The problem with angel investing is that such funds often invest in only the earliest financing rounds of the company's development. This works well if a software company has a fast track to revenue or growth and value can accrete quickly in the development cycle. But this becomes much more difficult with deep science companies, where investments of $25 million to $75 million are needed to get a company to commercialization. With each

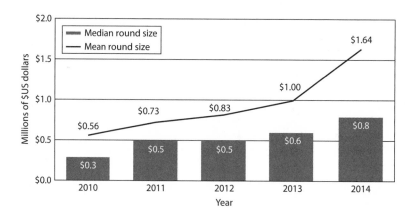

FIGURE 3.6 Median and mean round sizes of angel investors. The year 2014 saw both mean and median round size increase significantly. *Source*: Angel Resource Institute.

round, there is a risk that enough capital is not available to provide the next stage of growth. Without clear delineations of valuation accretion, the decade between 2001 and 2010 demonstrated that there was far too great a risk of being heavily diluted. Thus, angel investing, while growing as a portion of venture dollars, has actually decreased as a proportion of deep science investing.

Returning to the Payoff from R&D

The payoff from R&D spending is an important part of the process that drives innovation and economic dynamism over time. Absent a payoff from the billions of dollars spent on R&D annually, there is bound to be economic stagnation and diminishing prosperity over time. American venture capitalists have been integral to realizing the payoff to R&D in the postwar period in the United States by backing entrepreneurs with the requisite skills and resources to drive the commercialization process that brings new products to market.

It is noteworthy that the monetary resources commanded by entrepreneurs and the venture capitalists who support them year in and year out are quite small compared to other sources of funding in the innovation ecosystem. This fact makes the innovation accomplishments of entrepreneurs and venture capitalists all the more striking and remarkable and speaks to the nature of monetary funding (i.e., financial capital) as being a necessary, but insufficient, condition of successful innovation.[18]

As Apple cofounder Steve Jobs once observed, innovation is not about money per se. It is more about the people, how they are led, and how in tune innovators are with products that customers will love and purchase.[19] Entrepreneurs like Jobs, backed by venture capitalists, have played a vital role in generating significant and sometimes astronomical returns from R&D. That said, as the venture capital industry has evolved, there has been a meaningful shift in the innovation ecosystem. This shift has significant implications for the future innovation process and economic dynamism in the United States.

We can see this shift by analyzing the ecosystem for early-stage investments in deep science. Figure 3.7 represents how the ecosystem for early-stage investments in deep science looked during the 1980s and 1990s. One venture firm, in this case the Harris & Harris Group, invested approximately $180 million in early-stage deep science companies over the course of a decade. Another $85 million in nondilutive equity came from government sources, such as Small Business Innovation Research, ATP, and Defense Advanced Research Projects Agency funding. Two-thirds of the companies had corporate venture dollars or partnerships. An additional $1 billion in investment came from other venture capital firms.

Of the other venture capital partners who provided $1 billion of investment, the majority of these are no longer making active new investments in deep science as of 2014. Many are no longer in existence, and those that are have migrated to the software side of venture capital.

Venture Capital Diversity Breakdown

The migration of venture capital away from deep science over the past decade is a trend that merits further analysis and discussion, for it carries with it vital implications for the U.S. economy. Recall from the introduction the important finding in the economic literature that the amount of capital invested in R&D is less important than the payoff to that R&D in driving innovation, productivity, and economic growth. Historically, that payoff came from the translation of academically derived ideas into new businesses by company builders and corporations. However, over the past decade, we have seen a disquieting trend in the type of R&D that is being translated into economic growth in America.

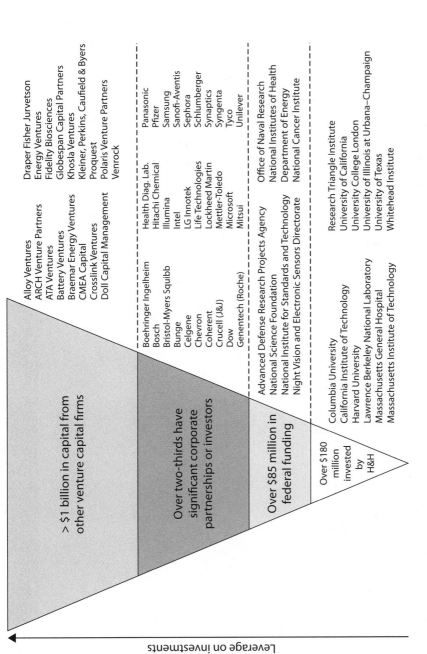

Leverage on investments

> $1 billion in capital from other venture capital firms

Alloy Ventures
ARCH Venture Partners
ATA Ventures
Battery Ventures
Braemar Energy Ventures
CMEA Capital
Crosslink Ventures
Doll Capital Management

Draper Fisher Jurvetson
Energy Ventures
Fidelity Biosciences
Globespan Capital Partners
Khosla Ventures
Kleiner, Perkins, Caufield & Byers
Proquest
Polaris Venture Partners
Venrock

Over two-thirds have significant corporate partnerships or investors

Boehringer Ingelheim
Bosch
Bristol-Myers Squibb
Bunge
Celgene
Chevron
Coherent
Crucell (J&J)
Dow
Genentech (Roche)

Health Diag. Lab.
Hitachi Chemical
Illumina
Intel
LG Innotek
Life Technologies
Lockheed Martin
Mettler-Toledo
Microsoft
Mitsui

Panasonic
Pfizer
Samsung
Sanofi-Aventis
Sephora
Schlumberger
Synaptics
Syngenta
Tyco
Unilever

Over $85 million in federal funding

Advanced Defense Research Projects Agency
National Science Foundation
National Institute for Standards and Technology
Night Vision and Electronic Sensors Directorate

Office of Naval Research
National Institutes of Health
Department of Energy
National Cancer Institute

Over $180 million invested by H&H

Columbia University
California Institute of Technology
Harvard University
Lawrence Berkeley National Laboratory
Massachusetts General Hospital
Massachusetts Institute of Technology

Research Triangle Institute
University of California
University College London
University of Illinois at Urbana–Champaign
University of Texas
Whitehead Institute

FIGURE 3.7 Traditional picture of the Harris & Harris Group, 2002–2010. All numbers are as of September 30, 2014, and include amounts invested and corporate partnerships since 2002, which is when the first of the current managing directors joined the firm. *Source:* Harris & Harris Group.

What we are seeing today is entrepreneurial financing that is significantly concentrated and focused on software and what can be called "creative commerce technologies" (e.g., media, entertainment, and financial services). As venture capital companies have turned to software-driven Internet and media deals, there has been a dearth of funding available for entrepreneurs seeking to foster the commercialization of innovations based in deep science. The data are striking and telling, with major ramifications for American productivity and economic dynamism.

4

Diversity Breakdown in Venture Investing

Software is eating the world.
—MARC ANDREESSEN

IN THE PRECEDING CHAPTERS, we have discussed the intertwin-
ing relationships between deep science, technological innovation,
and economic dynamism, in addition to the role venture capital
has traditionally played in fostering the development of early-
stage companies commercializing scientific technologies. In recent
years, there has been a disconcerting trend in venture investing
away from deep science investing toward software investing.
The lack of diversity in U.S. venture capital is currently unprecedented. It threatens to create instabilities in the economy that
will diminish the payoff to deep science R&D, a payoff that has
historically fostered economic dynamism and prosperity in the
United States.

Recent Trends in U.S. Venture Capital

From 1946 through the mid-1990s, much of venture capital
was focused on the translation of academic research. But in the

mid-1990s, the venture capital community began to focus on software investments, and most recently on digital, creative, and commerce technologies, all of which fit software investing models. According to a report by the State University of New York (SUNY) and New York State in 2012, 40 percent of all venture capital deals at the time flowed into information technology and 16 percent to creative and commerce deals. This trend left 44 percent of all deals for the life sciences and physical sciences.[1]

Federally funded academic R&D shows a very different pattern of investment—one of continued investment in the deep sciences. Over the past twenty years, the distribution of academic R&D has favored the deep sciences—life sciences and physical sciences. Currently, according to the SUNY report, life sciences represent approximately 57 percent of academic R&D, the physical sciences represent about 30 percent, down from the postwar years, and information technology and other areas together represent only about 13 percent.

The most striking trend in U.S. venture capital flow is the migration of venture investing to software and related deals. There has been a sharp increase in the percentage of venture deals concentrated in software and related companies over the past several years (figure 4.1). In 2014, over half (53 percent) of all U.S. venture capital was concentrated in the software and media/entertainment sectors.[2] Of the 4,356 U.S. venture capital deals that resulted in total investment of $48.3 billion in 2014, 1,799 deals and $19.8 billion were in the software segment, and another 481 deals and $5.7 billion were in media and entertainment.[3] These two segments combined attracted over $25 billion of venture capital in 2,280 deals in 2014.

Software has had the most venture dollars invested annually for the past five years. There has never been a time of such concentration in venture investment in software and related deals such as those in media and entertainment. It is noteworthy that Internet-specific companies captured $11.9 billion going into 1,005 rounds in 2014, marking the highest level of Internet-specific investments

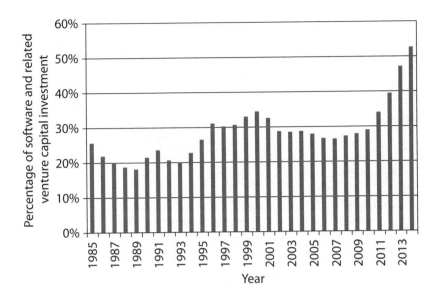

FIGURE 4.1 Investment in software as a share of total U.S. venture capital investment.

Note: Author's calculations, based on PricewaterhouseCoopers/National Venture Capital Association MoneyTree data.

since 2000. "Internet-specific" is defined by the National Venture Capital Association as "a discrete classification assigned to a company whose business model is fundamentally dependent on the Internet, regardless of the company's primary industry category." This classification is software intensive.

If government R&D is being focused predominantly on the deep sciences, but innovation capital is being focused predominantly on software, what is generating the payoff from government- and corporate-sponsored R&D? Without innovation being focused on R&D ideas and inventions, there will be less payoff from R&D and less value created from that research. The result will be that government funds invested in R&D will be less impactful, maybe even wasteful.

Once a mainstay of Silicon Valley, the portion of venture deals associated with backing innovative scientific technologies has

Table 4.1 U.S. Venture Capital Investments by Sector, 1985–2014

Segment	2014		1995		1985	
	Total investment (%)	Total deals (%)	Total investment (%)	Total deals (%)	Total investment (%)	Total deals (%)
Software and related	53	52	27	30	26	28
Deep science	25	25	44	41	56	51
Deep science, excluding biotech	14	14	33	32	51	46
Deep science/Silicon Valley legacy*	6	5	8	11	30	24

*Computers and peripherals, semiconductors, electronics and instruments
Source: Author's calculations, based on PricewaterhouseCoopers/National Venture Capital Association MoneyTree data.

steadily decreased over the past two decades. This migration away from deep science venture deals is illustrated in table 4.1.

Software and related deals represented 53 and 52 percent, respectively, of total U.S. venture capital investment and total number of deals in 2014. This is up from 27 and 30 percent, respectively, in 1995 and from 26 and 28 percent, respectively, in 1985. Hence, U.S. venture investing in software investments has doubled over the past two decades. Venture investing in the deep sciences has seen the opposite trend.[4] In 1985, deep science deals accounted for 56 percent and 51 percent, respectively, of total U.S. venture capital investment and deals. In 2014, deep science each accounted for 25 percent of total U.S. venture capital investment dollars and total number of deals.

Thus, we can see that deep science venture capital investments accounted for almost 60 percent of total U.S. venture capital dollars two decades ago. A large share of venture capital investing in deep science during the 1980s was linked to activities in Silicon Valley, whose roots are in deep science venture investing. In 2014, deep science venture deals had declined to 25 percent, and nearly half of the deep science venture deals in the United States were concentrated in biotechnology—a segment whose genesis is also associated with venture capital.

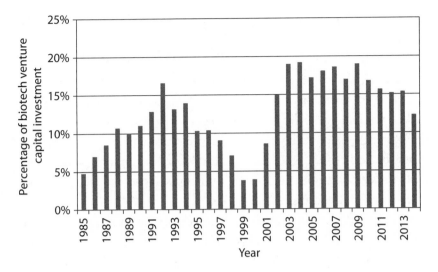

FIGURE 4.2 Investment in biotech companies as share of total venture capital investment. *Source*: Author's calculations, based on PricewaterhouseCoopers/ National Venture Capital Association MoneyTree data.

Figure 4.2 shows the share of total U.S. venture capital investment in the biotechnology segment. As can be seen, venture investing in biotechnology has been on a bit of a roller coaster ride since the early days of biotechnology in the mid-1980s, with a great deal of focus on new types of therapeutics.

Biotechnology investments ramped up in the early 1990s but then fell sharply as the decade progressed amid a proliferation of dotcom deals. The mapping of the human genome in the early 2000s—a significant deep science milestone with roots in biology, physics, chemistry, and computer science—and the collapse of the dotcom segment led to a revival of U.S. venture investing in biotechnology. The years 2001 through 2009 saw a resurgence of biotech venture capital investments, receiving nearly 20 percent of total U.S. venture capital investment during most of that period. In the past several years, however, investments in biotech have declined as a share of total U.S. venture capital investment from nearly 20 percent to 13 percent owing to a migration away from deep science toward software and social media.

In table 4.1 we see deep science investment, excluding biotech, falling from over 50 percent of total U.S. venture investment to 13 percent over the past two decades. Parsing the data further into what we call the Silicon Valley legacy segments—computer and peripherals, semiconductors, and electronics and instruments, sectors that one often associates with the birth of Silicon Valley—we see a severely diminished focus on deep science venture investing outside biotechnology.

Deals in these four deep science sectors accounted for only 6 percent of total U.S. venture capital investment in 2014, down sharply from 30 percent in 1985. Table 4.2 provides a more detailed breakdown of U.S. venture capital investment in deep science over the past two decades. The data show that, outside biotech and medical devices, there is currently very little deep science U.S. venture capital activity. As deep science venture entrepreneur

Table 4.2 U.S. Venture Capital Deals by Sector, 1985–2014

Segment	2014		1995		1985	
	Total investment (%)	Total deals (%)	Total investment (%)	Total deals (%)	Total investment (%)	Total deals (%)
Software segments	53	52	27	30	26	28
Software	41	41	15	23	22	24
Media and entertainment	12	11	12	7	4	4
Deep science segments	25	25	44	41	56	51
Biotech	12	11	10	9	5	5
Medical devices	6	7	8	9	7	10
Computers and peripherals	3	1	4	5	16	12
Semiconductors	2	2	3	3	9	6
Electronics and instruments	1	1	2	3	4	5
Networking equipment	1	1	5	4	8	6
Telecommunications	1	1	12	7	6	6
Industrial/energy	5	6	7	7	7	9

Source: Author's calculations, based on PricewaterhouseCoopers/National Venture Capital Association MoneyTree data.

and Nantero cofounder and CEO Greg Schmergel once remarked to us, Silicon Valley is not interested in silicon anymore.

Entrepreneurs engaged in commercializing scientific technologies—especially those outside biotechnology and medical devices—are being shunned by Silicon Valley venture capital funds and forced to find new sources of seed and growth capital, both from within and outside the United States. Increasingly in conversations with entrepreneurs engaged in commercializing deep science outside biotechnology, we hear of the need to travel to Asia and Europe to secure early-stage investments, as entrepreneurs engaged in commercializing innovative deep science technologies are finding greater interest from investors overseas. Investors outside the United States tend to have the investing time horizons (i.e., the periods of time over which money is invested without expecting a return) needed for the development of scientific technologies.

The migration of deep science venture investment activity away from Silicon Valley and the United States clearly presents some long-term challenges for the United States given the high degree of economic dynamism associated with deep science venture investing. The increased concentration of venture capital in software investments threatens to undermine an important source of long-term American innovation—employment, income, and productivity growth that propels living standards ever higher over time. Trends in U.S. venture investing are pointing to what complexity scientists refer to as a "diversity breakdown."

Diversity Breakdown in Venture Investing

Although total U.S. venture capital investment in 2014 fell far short of the peak of $105 billion seen in 2000 during the height of the dotcom era, few voices today are expressing concern about the traditional venture capital model being broken and no longer relevant. Media headlines touting U.S. venture capital, such as

"VC Fundraising Is on Best Pace Since 2000" and "QI Investment in VC Funds Highest in a Decade," have been positive and upbeat in recent years amid a greater concentration of deal flow into software investments. The recovery in the venture capital business since the early 2000s dotcom bust is most welcome and is certain to be recognized as a positive development given the important role venture investing plays in fostering technological innovation and economic dynamism.

What is concerning about the current venture capital environment today, however, is the increasing concentration of investments in software and the resulting migration away from deep science. This shift is unprecedented in the decades of modern venture capital. The trend underway, should it strengthen and persist, has the makings of what complexity scientists refer to as a "diversity breakdown." Diversity breakdowns occur in complex systems when the structure of the system evolves into a narrow group of agents or elements. Complexity scientists note that robust complex systems contain a diversity of agents or elements, and it is this diversity that maintains balance and stability within the system.

In other words, the dynamics of a complex system are a function of the diversity of agents within the system. As the diversity of agents within a complex system such as the stock market declines—through investor focus on one or a few segments within the stock market—the system becomes unstable and prone to large shifts. During times of dwindling diversity, what seem like minor disturbances can produce large effects. Complexity scientists note that most of the changes in a complex system take place through catastrophic, "black swan" events, rather than by following a smooth, gradual path.[5] This insight helps to explain stock market crashes and other financial market meltdowns—financial market events that traditional economic models have difficulty explaining.

With respect to complex systems such as the U.S. economy or stock markets, diversity is the default assumption made in conventional theoretical models. A diversity of investor types within the stock market produces a state of efficiency that leads to equilibrium. A stock market is composed of an ecosystem of agents or investors. Some agents have short-term time horizons (e.g., traders), some have longer-term horizons (e.g., so-called value investors), others focus exclusively on small capitalization stocks, and others concentrate on companies with relatively large market capitalization (e.g., the companies in the Dow Jones Industrials Index). Within the ecosystem of the stock market, there are agents who trade or invest in various segments of the market (e.g., energy, technology, consumer products). Taken together, these agents represent what could be termed a "full ecosystem," akin to all the animals, insects, and plants that make up the ecosystem of a rainforest. The existence of a diversity of investor types is generally sufficient to ensure there is no systemic way to beat the overall stock market, and it is assumed that this diversity of investors produces an efficient market.[6] Market inefficiencies can arise when the assumption of diversity is violated—that is, when there are diversity breakdowns.

Diversity breakdowns in financial markets occur when investors engage in lemming-like herding behavior. Herding in financial markets occurs when investors make similar investment decisions based on observations of the behavior of others, independent of their own information and skills. Herding behavior among investors leads to manias and bubbles in financial markets. Studies of stock market booms and busts are linked to manias, panics, and what Charles Mackay called "extraordinary popular delusions and the madness of crowds."[7] Manias and bubbles—extraordinary delusions and crowd madness—are associated with heightened concentrations of narrow investment in a sector or industry segment. Heightened concentrations of investment in a particular sector or segment is the hallmark of a diversity breakdown. During such periods, valuations are pushed higher and higher, and rising valuation attracts more investors. Over time, herding behavior pushes valuations to extreme levels that are difficult to justify using conventional financial metrics.

Valuations continue to rise as more capital flows into the sector. At some point, diversity in the stock market breaks down, signaling a major inflection point in the market. Such inflection points are followed by a major change in investor behavior that works to restore diversity to the market.

Investors may remember what happened after the U.S. stock market's diversity breakdown of the early 2000s. The period leading up to the breakdown was marked by an increased concentration in dotcom companies. As concentration in the dotcom segment increased through investor herding, Internet mania ensued in the equity market. As diversity dwindled in the stock market through an ever-greater concentration of investments in the dotcom segment, valuations rose sharply to levels that were difficult to reconcile with fundamental accounting and traditional valuation metrics.

The diversity breakdown of the early 2000s signaled the end of dotcom mania, Soon afterwards, valuations of dotcom stocks collapsed, and it was only then that diversity began to be restored to the stock market. It should be noted that the restoration of diversity within the market was accelerated by major policy changes in the form of new corporate and capital market regulations.

Through the study of complexity, we see the importance of agent diversity within a complex system. Complexity science sheds a great deal of light on the dynamic behavior of complex systems such as financial markets. Complexity science informs us of the importance of diversity to system robustness and stability. Complex systems become prone to instability and major shifts during times of diversity breakdown.

The amount of venture capital dollars flowing into software investment today could signal the beginnings of a diversity breakdown. If such a trend persists, one can confidently predict the manifestation of a series of dynamics leading to high and rising valuations of venture-backed software deals and a further migration away from deep science venture deals.

Some astute venture investors have expressed concern about the potential for a diversity breakdown in venture investing related to the increasing concentration in software investments. One such investor is Steve Blank, a serial entrepreneur, author, and Stanford Graduate School Business professor of entrepreneurship. In the spring of 2012, Blank detected a major shift by venture capitalists in Silicon Valley away from deep science toward social media software investments.[8] The rise of social media has captivated a large and growing share of venture interest in Silicon Valley. In the current venture environment, notes Blank, venture capital firms in the Valley that would have looked to back deep science investments are now interested only in whether something is associated with a smartphone or tablet.

Blank notes that companies such as Facebook—one of the poster children of the social media rise—have capitalized on market forces on a scale never seen in the history of commerce:

> For the first time, startups can today think about a total available market in the billions of users (smartphones, tablets, personal computers, etc.) and aim for hundreds of millions of customers. Second, social needs previously done face to face (friends, entertainment, communication, dating, gambling, etc.) are now moving to a computing device. And those customers may be using their devices/apps continuously. This intersection of a customer base of billions of people with applications that are used/needed 24/7 never existed before.[9]

The potential revenue and profits from these users, or advertisers who want to reach them, is unprecedented, and the speed of scale of the winning companies can be astonishingly fast.

Blank further notes that the Facebook initial public offering reinforced the new calculus for venture investors. In the past, a talented Silicon Valley venture capital fund could make $100 million on an investment in five to seven years. In the current environment, social media startups can return hundreds of millions or even billions of dollars in less than three years. Witness the meteoric rise of software apps such as Uber and other so-called on-demand and mobile Internet companies.

Venture funding in the on-demand mobile services sector is driving an increasing concentration of venture investing in software today. Financing trends in this segment have been nothing short of explosive, with many prominent Silicon Valley venture capital funds aggressively investing in the sector. The star of this segment is Uber, which has attracted over $5.5 billion of venture funding during the past five years, according to CB Insights. As of 2015, the private valuation of Uber was reported to be in the neighborhood of $50 billion, a valuation in excess of established companies that have been in business for a far longer time. In the span of five years, Uber has gone from a privately held company with a post-money valuation of $60 million to a company seeking to raise additional capital at a $50 billion valuation (box 4.1). The nearly ten-fold increase in Uber's valuation since 2011 has been nothing short of astonishing when you consider that the current market capitalization of General Motors is around $58 billion.

BOX 4.1
Uber Venture Financing, 2011 to 2015

February 2011: $11 million raised at a reported post-money valuation of $60 million
November 2011: $37.5 million raise at a reported post-money valuation of $330 million
August 22, 2013: $350 million raised at a $3.5 billion valuation
June 6, 2014: $1.2 billion raised at a record-breaking $17 billion valuation
December 4, 2014: $1.2 billion raised at a $40 billion valuation
May 9, 2015: Seeking $1.5 billion to $2 billion at $50 billion valuation

Source: VentureBeat.

Financing for other on-demand startups has not been as pronounced as Uber's but nevertheless has been quite robust, especially relative to the financing associated with deep science companies. According to the proprietary database of CB Insights, at the time of this writing Lyft had raised over $862 million, Airbnb over $794 million, Instacart $275 million, Eventbrite nearly $200 million, Thumbtack nearly $150 million, and FreshDirect nearly $110 million. In the span of five years, the number of venture capital investors that have done a deal in on-demand mobile services has gone from less than twenty to nearly two hundred.

According to analysis by Internet advisory firm Digi-Capital, there were seventy-nine companies that fulfill the "unicorn" criterion: companies with valuations in excess of one billion dollars. The combined market value of these companies in the first quarter of 2015 was estimated to be a massive $575 billion, with Facebook alone accounting for nearly 40 percent ($220 billion) of the total value. United States–based mobile Internet software companies represent the lion's share of the value of unicorns globally, accounting for three-quarters of that value.[10]

Amid the exuberance in venture activity in the software segment associated with the ascent of Uber and mobile Internet unicorns, some experienced venture investors are beginning to express concern about the valuations being commanded by startups. Respected Union Square Ventures software investor Fred Wilson observed in a May 2015 blog post that in the current environment, it is difficult to make an investment in the software sector because "the math doesn't work."[11] He notes that when valuing a venture-stage opportunity, one has to imagine that the product can scale to be used by many more people or companies, or both, than are using it currently. It is in scaling up a business that companies achieve a state of profitability that fosters additional growth and penetration in the marketplace that leads to success and prosperity.

For this exercise, says Wilson, a venture investor needs to analyze the product, the road map, and the use cases to ensure that what one is imagining is possible and not delusional. One also needs to figure out what an annual revenue per user might be and apply that to the potential size of the market. Then, one needs to study the economics of the business and figure out how much of that potential revenue might flow to the bottom line. Finally, Wilson notes, a venture investor needs to figure out how the market might value the cash flow of the business. After performing such analysis, the venture investor needs to discount the value of the cash flow by three times, five times, or even ten times. The discount factor reflects the risk inherent in early-stage ventures associated with not knowing what will actually transpire in the future.

Analyzing venture deals in the software sector today with the process outlined earlier makes it difficult to invest, states Wilson. The numbers just do not add up. He suspects that many venture investors are not doing this type of work today, which would help explain the frothy valuations being commanded by startups in the software segment at the present time. One might say that investors are suffering from "Uber mania" (see box 4.1).

During times of diversity breakdowns in investing, valuations are pushed to higher and higher levels. At some point that is difficult to predict, valuations become unsustainable in light of economic fundamentals. We can view Fred Wilson's comments as an indication that, while the days of frothy valuations may continue in the software sector, there is a growing reluctance among some experienced venture investors to participate in new software enterprises.

There are several reasons for the pronounced shift in venture capital away from deep science deals toward software investments. At its essence, venture investing, like all forms of investing, is an exercise in Bayesian thinking and statistical distributions. Authors Benoit Mandelbrot, Michael Mauboussin, and Nassim Taleb have done much recently to articulate the statistical nature of investing. Unlike many other forms of investing, however, venture capital investing operates on a long-tail distribution. In investing, this term is applied to frequency distributions that often form power laws and are thus long-tailed distributions in a statistical sense. There are very few investment winners, and most investments lose money for venture investors. Returns are dominated by a few "home-run" investments.

Because early-stage venture capitalists are investing in new technologies, often in unproven industries and with new management teams, there is more luck than skill involved in the ultimate success of an investment. In his book *The Success Equation*, Michael Mauboussin describes the skill–luck continuum. On the far right are activities that rely purely on skill and are not influenced by luck. On the far left are activities that depend on luck and involve no skill. Roulette and the lottery would fall strongly on the luck side of the continuum, whereas a sport like tennis would fall very far on the skill side of the continuum.[12] The earliest financing rounds of venture capital fall far closer to the luck side of the continuum than to the skill side.

Mauboussin notes that if you have an activity where the results are primarily the result of skill, a small sample size can be used to draw reasonable conclusions. It does not take long to identify the best tennis player or the fastest runner. However, as you move further left on the continuum toward the luck extreme, you need larger and larger samples to understand the separate contributions of skill and luck. Mauboussin discusses a game of poker in which a lucky amateur may beat a pro in a few hands but in which the pro's edge becomes clearer as more hands are played.

Playing poker, like venture investing, is an exercise in using Bayesian statistics. The goal, as Charles Duhigg explains masterfully in *Smarter, Faster, Better*, in a chapter titled "Decision Making: Forecasting the Future (and Winning at Poker) with Bayesian Psychology," is to make predictions, imagine alternative futures, and then calculate which alternative futures are most likely to come true.[13] Duhigg notes that to be among the elite in poker, bets need to be ways of asking other players questions and using the answers to predict the future a little more accurately than the other players in the game. Chips are used to gather information faster than others in the game.

Poker is a game of using the insights from the answers of questions asked to forecast more likely futures. You never know with 100 percent certainty how the game will turn out, but the more potential futures envisioned, the more you learn which assumptions are certain and which are not, and the better your odds of making a better decision next time. This is Bayesian thinking: using new information to improve assumptions as you move forward. Poker is in essence thinking probabilistically to embrace the uncertainty inherent in the game.

Blackjack is a game of gambling located more toward the luck end of the skill–luck continuum than poker. Using Bayesian thinking in blackjack requires something the casinos will not allow: counting cards. There is a story about Jeff Ma and a team of blackjack players from MIT, made famous in Ben Mezrich's book, *Bringing Down the House*. The MIT students would begin the evening at the casino counting cards at a number of different tables to determine which tables had more attractive odds of winning. The players bet small amounts of money at this stage. They were playing to determine if there was a relatively larger number of higher cards remaining in the shoe: the more high cards, the greater the chance that the player would win a hand. When players found an attractive table, teammates would join and begin placing large bets in order to win as much money as possible. As the odds changed in their favor over time, the bets had a better chance of making them winners.

High-quality venture investors can, in effect, "count cards" and thus change the statistical distribution of success with each new round of investment they make. For instance, if I am playing blackjack and I am counting cards in the deck, at each subsequent deal, I can recognize the statistical odds changing, potentially in my favor for getting the cards I need to reach twenty-one. By placing larger bets when the odds have increased in my favor,

I increase my chances of winning. I change my statistical distribution for success.

Venture investing is similar. During the first rounds of investment, success is dominated predominantly by luck. Investors want to invest as little as possible to gain access to the game, and they want to reserve much greater capital for the future. Some companies will be lemons, and no further investment will be warranted. However, some companies will demonstrate prescient management or win a crucial supply deal with an important customer, as in the case of Tandem Computers with Citibank (see chapter 3). This may change the statistical distribution of success, and as an investment's probability for success begins to change so that the odds favor success, a venture investor wants to invest more money into the company.

The statistical nature of venture capital investing and the ability to effectively "count cards" work extremely well for software investments. Because many of these innovations tend to be software, rather than hardware, focused, large sums of capital to jump-start the idea are not required as much as they are in deep science investing. With a small sum of capital, often now provided by angel investors, an entrepreneur can write the code to demonstrate the viability of a new product. Then, with the help of the Internet, that same entrepreneur can see if there are people interested in the product. Again, the capital outlay is still small. Once it is determined that there are many customers for the product or service, the larger sums required to grow the market can be invested by institutional venture capitalists. Entrepreneurs can get further on less capital, and investors can have far more "cards counted" before they have to invest large sums of capital for future growth.

This is a very effective model for venture capital investing, and because of this, software-related ventures have attracted a great portion of the capital that has historically invested in a more diversified way in the venture capital market.

The Beginning of the End of a Venture Era

The calculus surrounding software investments today is working to siphon venture capital away from deep science startups. Blank describes this calculus bluntly: "If venture investors have a choice of investing in a blockbuster cancer drug that will pay them nothing for fifteen years or a social media application that can go big in a few years, which deal do you think they are going to choose?"[14] He goes on to note that venture capital firms are phasing out their traditional science divisions.

Further, venture investors are becoming more reluctant to fund deep science clean (or green) technology. As clean tech venture capital funds have painfully learned, trying to scale innovative clean technology past demonstration plants to an industrial scale takes capital and time often in excess of the resources of venture capital. Compared to iOS and Android apps, says Blank, all other deep science venture investing is difficult because the returns take much longer to materialize.

Because of the size of the market opportunity and the nature of the applications, the returns on social media and software-related investments are quick and potentially huge. Blank observes that new venture capital funds focused on both the early and late stages of social media development have transformed the venture capital landscape. This transformation shows up clearly in the venture capital flow data, which show an increasing concentration in software and a migration away from other deep science deals, like those related to green energy.

Moreover, as venture investor Mark Suster notes, changes in the software industry over the past five years are having a significant impact on the venture capital business.[15] Suster has stated that when he built his first software company in 1999, it cost $2.5 million in infrastructure to get started and another $2.5 million in team costs to code, launch, manage, market, and sell the software. A typical

"A round" of venture capital for software companies back then was in the neighborhood of $5 million to $10 million. Suster points out that the trend toward open-source software and horizontal computing has worked to significantly lower infrastructure costs to the point where it has become almost free. A move to horizontal computing has alleviated the need to buy expensive UNIX servers and multiple machines to handle redundancy.

Another major shift in the software industry, beyond the move toward open-source computing, is associated with the "open cloud." Open-cloud services are provided solely for the economic purpose of building a cloud-based business, as opposed to the "platform cloud," where certain service providers offer cloud services wrapped around their core products (e.g., Salesforce.com, Cloud Foundry, Microsoft Azure). Suster observes that entrepreneurs wanting to build independent, high-growth, venture capital–backed startups today can build the company on a truly open cloud (i.e., a cloud that is not associated with cloud services wrapped around a core product).

Where open-source computing gave a software startup a 90 percent reduction in software costs, open-cloud services, such as those provided by Amazon.com, provide software startups a whopping 90 percent reduction in total operating costs. Suster notes that Amazon.com has allowed 22-year-old tech developers to launch companies without even raising capital. The open cloud has profoundly sped up the pace of innovation because, in addition to not having to raise capital to start, entrepreneurs no longer have to wait for hosting to be set up, servers to arrive, or software to be provisioned.

We see that not only has the proliferation of smartphones, tablets, and social media altered the venture investing landscape over the past several years, but there has also been a significant shift in the software industry that has greatly diminished the infrastructure costs needed to start up a software business today. The Amazon open cloud revolution of startup innovation, says Suster, has led to a massive increase in the aggregate number of software startups. This, in turn, has fueled incubation programs such as Y Combinator, TechStars, 500 Startups and many more to help early-stage teams launch businesses led by mostly technical founders who are getting coaching from seasoned management teams.

Additionally, it is much easier to get distribution than it was in the pre-Facebook, pre-iPhone world. Suster notes that it is not uncommon to see a team out of Utah, Texas, or, for that matter, Finland with eight to ten developers building iPhone apps that get tens of millions of downloads and receive hundreds of millions of monthly page views.

What's more, changes in the software business are fundamentally altering the structure of the traditional venture capital business toward what Suster refers to as "micro VCs." The typical "A round" size of venture capital in the late 1990s of $5 million to $10 million has shrunk significantly to between $250,000 and $500 thousand. Shifts in the software industry led by Amazon.com have led to the birth of the micro VC and a blurring of the lines between traditional venture capital funds and later-stage investment firms. Additionally, the U.S. venture investing landscape is increasingly being populated by angels, incubators, mutual funds, and hedge funds.

Blank and Suster note that the heightened activity in the software industry and the advent of social media amid a proliferation of billions of electronic devices has resulted in a change in the venture investing landscape toward software and away from deep science deals. That said, what is great for making tons of money may not be the same as what is great for innovation in our country, as Blank has astutely observed. For Silicon Valley, the investor flight to social media, says Blank, marks the beginning of the end of the era of venture capital–backed big ideas in science and technology.

This statement, if true, carries with it significant implications for the future of American economic dynamism. As we have seen in previous chapters, big ideas in science and technology have been a prominent and growing source of economic dynamism in the national and global economies. And over the past half-century, venture capital, and Silicon Valley in particular, has played an integral role in fostering the development of deep science technology. As we have noted, venture capital is well suited to the task of commercializing deep science innovations. If we are truly observing the end of the era of venture capital–backed big ideas in science and technology owing to the rise of social media and software, as Blank observers, the economic consequences of such a shift will be profound.

For decades, notes Blank, the unwritten manifesto for Silicon Valley venture capital funds has been a quote from John F. Kennedy: "We choose to invest in ideas, not because they are easy, but because they are hard, because that goal will serve to organize and measure the best of our energies and skills, because that challenge is one that we are willing to accept, one we are unwilling to postpone, and one which we intend to win."[16]

Today, this unwritten manifesto is being challenged by a confluence of factors that are conspiring to produce a diversity breakdown in venture investing and a migration away from deep science. Major factors, such as the relatively favorable calculus of investing in software startups in an age of social media relative to the calculus of investing deep science–enabled technology, threaten to put an end to venture capital–backed big ideas in science and technology. One wonders how a deep science inventor like Nikola Tesla would be treated today in Silicon Valley.

A final example of the breakdown in the diversity of venture investments shows up in the recent phenomenon of the "unicorn." This term, coined by Silicon Valley venture investor Aileen Lee, now of Cowboy Ventures, is used to describe private venture–backed firms that have reached a venture valuation of at least $1 billion. This phenomenon was relatively rare until a few years ago, as most companies that reached market valuations of $1 billion before then did so in the public markets. But with more companies staying private longer, and with public money flowing into the private market chasing software deals, the number of unicorns ballooned to over 132 companies by November 2015.[17]

Of these 132 unicorns, only ten could be considered deep science companies, and of these ten, four are biotech firms: this is less than 10 percent of the unicorn population. Consumer-oriented companies drive the majority of value of unicorns, being the most common type of company and having the highest average value per company. In terms of business models, e-commerce companies are driving the majority of unicorn value.[18]

Whither Deep Science Venture Investing?

It has never been easier to start, fund, and grow a software company. On the other hand, it has never been so difficult to start, fund, and grow a deep science enterprise outside the biotechnology segment than it is today. While software companies can ramp up and scale business at a relatively fast rate, as we have seen with Facebook, Uber, and others, the process associated with commercializing transformative technologies based on deep science is anything but fast.

While software companies can write an application in a matter of weeks and launch it in the market to a wide audience, it often takes several years of intense R&D work to ready a deep science product for commercialization. Additionally, depending on the disruptive nature of a deep science technology, it can take several more years to begin to penetrate the market in a meaningful way—that is, in a way that generates sufficient cash to sustain the company without further participation from investors. The lengthy product development times coupled with the extended time it takes to penetrate the market puts deep science ventures at a major disadvantage relative to startups in the software segment.

Richard Thaler, Amos Tversky, Daniel Kahneman, and Alan Schwartz's paper, "The Effect of Myopia and Loss Aversion on Risk Taking: An Experimental Test," provides a great explanation as to why deep science investing has suffered at the hands of software investing. In addition to explaining the difference between the expected-utility theory and prospect theory and the equity–risk premium, the authors note the following important concept: "The attractiveness of the risky asset depends on the time horizon of the investor. An investor who is prepared to wait a long time before evaluating the outcome of the investment as a gain or a loss will find the risky asset more attractive than another investor who expects to evaluate the outcome soon."[19]

It is this topic that we will pursue further in chapter 5. The migration of venture capital toward software is understandable in light of the investment realities in the marketplace today.

Fostering Diversity in Venture Investing

FOLLOWING WORLD WAR II, deep science R&D and mechanisms for realizing a payoff from that R&D blossomed through the innovations of American venture capital. However, after some vibrant decades between the 1960s and 2000s, multiple factors have worked in concert to lead to a breakdown in the diversity of the types of investment in which venture capital resources are focused. This has not necessarily been bad in the short term for venture capital, but the loss of diversity has impacted the payoff from deep science R&D. This diversity breakdown has also coincided with a slowdown in productivity growth and waning economic dynamism.

In this chapter, we focus on how changes in investing time frames and structural changes in the public markets have together further disadvantaged deep science investing. We then begin to explore possible mechanisms for returning deep science investing to its rightful place as a contributor to economic dynamism.

Time Scales for Measuring Success

Before examining a major cause of the breakdown in deep science investing in the early 2000s, it is important to first discuss why the time line by which we measure investment success is an important determinant when working with deep science innovation. The classic depiction of value accretion in angel and venture investing is a valuation staircase ascending linearly up and to the right (figure 5.1).

The idea is that with each new advancement made by an entrepreneurial company, value is recognized in the market by a step up in the valuation that this company receives from the investing marketplace. After a series of advancements, the company is noticeably more valuable than it was at the beginning. This means that an investor contributing at time zero will be able to recognize a value increase at each successive period. In a real marketplace, an investor could choose to sell at the first inflection point, or at

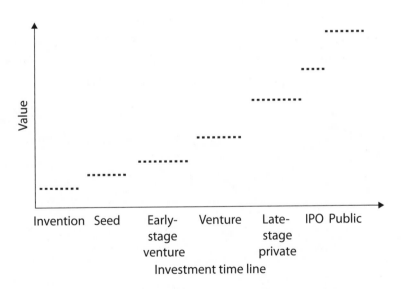

FIGURE 5.1 Staircase value appreciation. *Source*: Harris & Harris Group.

some point in the future, getting a return on his or her investment over whatever period of time the investor would like to hold the company. The market for innovation investing works effectively if this is the case. Unfortunately, for most new ventures and often for deep science investing, this is not the case.

Why is this? In their book, *The Rainforest: The Secret to Building the Next Silicon Valley*, Victor Hwang and Greg Horowitt point out that the risk premium on the continuum of capital for venture capital investments often does not follow classical economic theory.[1] The right-hand graph in figure 5.2 illustrates the reality of the investment return in early-stage deep science investing compared with classical economic theory, depicted in the left-hand graph. For most deep science investments, the cost curve is above the return curve until you get to a later stage in the company's development.

Over the past decade, especially in the electronics and semi-conductor markets, we have witnessed many deep science companies that have attained over $100 million in revenue, a success by many standards. Nonetheless, it has often taken five to ten years to obtain revenue, at which point it has taken another five years to grow that revenue to $100 million. Often, it has taken over $200 million to $300 million of invested capital to build

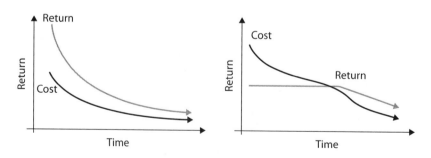

FIGURE 5.2 Classical economic theory (*left*) versus venture reality (*right*). *Source*: Victor W. Hwang and Greg Horowitt, *The Rainforest: The Secret to Building the Next Silicon Valley* (Los Altos Hills, CA: Regenwald, 2012).

these companies from their inception. Until these companies attain two to three times $100 million in current revenue through future organic growth, the cost of the investment exceeds any return.

Additionally, because of the amount of capital and the time required to bring deep science investment to commercialization, many variables may have an impact on value other than just the company's progress over that period of time. Macroeconomic trends and investor demand can change during the time periods required to bring deep science innovation to commercialization. Because a great amount of capital is needed during manufacturing and scale-up, it is often difficult to measure whether value is increasing during certain periods of time in the development cycle.

Investors in deep science also realize that not only is there a period of time during which investment costs exceed returns across the investing cycle—which is antithetical to classical economics—but that, graphically, value accretion when accounting for cost actually looks more like an asymptotic curve than a staircase. And our real-world experience is that value accretion is sometimes negative or downward-sloping before continuing its upward ascension, even in companies that are ultimately successful.

The curve shown in figure 5.3 does not resemble the linear progression of the staircase described earlier. In reality, the curve for deep science investing is often asymptotic, remaining very flat throughout the first years and then increasing as the market it creates or modifies begins to show acceptance. There is a long period of relative flatness before value begins to increase rapidly. In other words, value may stay relatively flat for the first five to ten years of some deep science investments before increasing rapidly over the following five to ten years as the innovation begins to penetrate and change the marketplace. What does this new, realistic curve for value accretion mean for measuring success? It means that success is quite dependent on the time frame over which you are measuring. If your holding period for an investment is only five years, you may have successful companies that have not yet

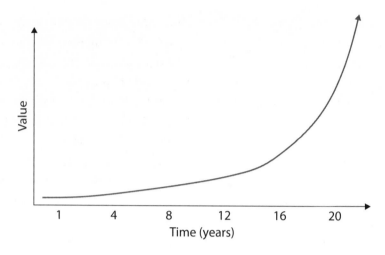

FIGURE 5.3 Realistic value creation curve for deep science investing. *Source:* Harris & Harris Group.

realized a value point in excess of the cost of that investment, especially when accounting for risk. However, if your holding period is fifteen to twenty years, the risk–return profile could be very rewarding, even after accounting for the time value of money. A parabolic value accretion curve for deep science investing creates a very different value proposition, dependent on time, than does a linear staircase depiction of return (figure 5.4).

Software investments, on the other hand, often take just months to a year to develop the technology. Very rapidly, through the Internet, it can be determined if there is a large market for the new product. Then, capital can be invested to fuel revenue growth and customer adoption. This entire process may take just five to seven years, rather than the ten to twenty years characteristic of deep science hardware technology investments.

Most venture funds are ten-year partnerships. Most of the new investments are made during the first three years of the fund, with the remaining seven years devoted to further investing and harvesting the earlier investments. A traditional venture capital fund has

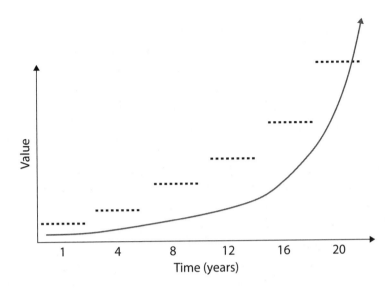

FIGURE 5.4 Deep science value creation versus the classic staircase. *Source*: Harris & Harris Group.

to invest and exit, ideally realizing a return commensurate with the risk that early-stage investments warrant in a period of seven to ten years. This works for software investment deals. However, this is very difficult to do for deep science investments in the current environment. But if you were to change the measuring period to twenty years, many of deep science investments could generate as large a return, or larger, even when adjusting for time.

The path to value creation for transformative deep science products extends over many years, often decades. The innovation associated with deep science is often displayed graphically by an archetypal pattern that is reflected in technology stock prices following IPO. The pattern is one of initial investor ebullience that drives the stock price above the offering price. Later on, there is a wave of investor selling that pushes the stock price down—sometimes to below the IPO price. This is followed by a long period of consolidation and then a resurgence of value creation that is reflected by a rising stock price over time. This archetypal pattern can be observed in the stock prices of Intel and Amgen, both long-time innovators in deep science.

Intel Corporation was founded in 1968 and completed an IPO in 1971, raising $6.8 million at $23.50 per share. (Note that that $6.8 million is equivalent to around $40 million today, which would qualify as a microcapitalization IPO). The company's stock price displays a pattern often seen with companies innovating in deep science (figure 5.5). Initial investor euphoria results in a sharp increase in the stock price, which is followed by a selloff. Consolidation ensues and gives way to a period of sustained value creation reflected by a rising stock price over time. In Intel's case, the development of the microprocessor—a truly transformative product—ushered in a computing revolution that continues today.

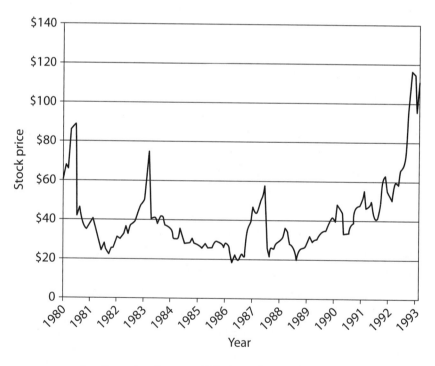

FIGURE 5.5 Intel's stock prices post-IPO. Data from YahooFinance.com.

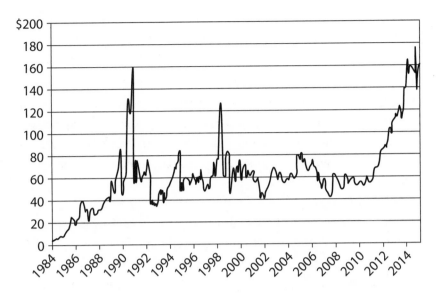

FIGURE 5.6 Amgen Inc.'s stock prices post-IPO. Data from YahooFinance.com.

Figure 5.6 shows the post-IPO stock price performance for Amgen Inc., a company founded in 1980 that completed its IPO in 1983 with nearly $40 million (approximately $96 million in 2016 currency). The discovery of the novel therapeutic Epogen in 1983 set the stage for Amgen's IPO and created quite a buzz in the stock market, as indicated by the significant rise in the company's stock price post-IPO. Periods of consolidation followed by the announcement of new discoveries propelled Amgen's stock price higher over time.

As the histories of Intel and Amgen show, the path to deep science value creation extends over a relatively long time span—one measured not in quarters or years, but decades. Within that long time frame, a great deal of volatility is reflected in the stock prices. Stock price volatility is part and parcel of the deep science innovation process, just as business cycles are for long-term macroeconomic growth and development.

Deep Science: The Great Relay Race

To return deep science investing to anywhere near its historical participation in the innovation ecosystem, and to have a vibrant innovation economy, there must be a way for each successive investor in deep science investing to pass the baton to the next participant in what becomes an extended relay race. Just as in a running relay, innovation progresses most efficiently when an early investor can "pass the baton" to the next investor, and when this process can be repeated multiple times. "Passing the baton" can mean selling your position to a later-stage investor or simply having a new investor take over the funding while existing investors hold on to their ownership of the investment. Innovation runs at a faster and more efficient pace in a relay than if one "runner" must complete the entire "race" alone.

Historically, the United States has led the innovation relay. Deep science innovation was successful because a network of investors emerged that could move development forward over a long period of time in a relay race fashion. Specifically, the engine of deep science technology began with strong government support and the belief that financing technological advancement would lead to economic growth. Entrepreneurs then developed scientific breakthroughs with access to angel investors and venture investors. This investment moved deep science from ideas to products. The angel and venture investors could then pass the baton to later-stage private investors, corporate entities, and early public investors. The American stock market had developed to support small, microcapitalization companies. These public markets also began to provide the first transparent look at the business of developing innovative companies. Then, within the public markets, innovative companies could grow. Access to capital became more widely available as companies increased in size and scale.

This process was effective from the 1950s through 2000 in the United States, making American scientific discovery the poster child for global success. However, in 2000, as a consequence of many regulatory changes that began in the 1970s, the public market for smaller innovative technology companies seized up. Without this critical member of the relay team, over the next decade and a half, the relay race broke down. Earlier participants found it more difficult to participate if there was no one to follow them into the innovation process. Without the ensuing participant to "pass the baton to," the relay race no longer worked and deep science investments began to sputter. No one is willing to run the first leg of a relay race if later-stage participants are not there to complete the race.

Losing Our Anchor Runner (2000 to 2016)

After World War II, the U.S. public markets became the "anchor" investor on the deep science innovation continuum. Angel investors and family investors passed the baton to venture investors who passed the baton to late-stage investors who took the companies public as IPOs. As discussed in chapter 2, many of ARD's investments were quickly listed as public companies on smaller exchanges, and grew on these exchanges. When Intel went public in 1971 it was not the behemoth it is today, but rather a tiny semiconductor company raising approximately $40 million (in today's dollars) in a public offering. The same is true with Amgen and Genentech. Today, they are Fortune 500 companies built over thirty- to forty-year time periods with relatively small public market debuts.

Would it be possible to take a tiny company like the Intel of 1971 public today? The answer could very well be no.

In their 2011 paper, "Why Are IPOs in the ICU?" David Weild and Edward Kim discuss the changing fates of IPOs.[2] In the 1990s, small IPOs raising less than $50 million were plentiful (between 60 and 80 percent of all IPOs) and dominated the IPO market. The authors argue that the dotcom bubble did not cause an increase in the abundance of IPOs, but rather dramatically expanded the size of preexisting IPOs during this bubble period. In the years following the dotcom crash, the number of large IPOs did not decrease dramatically, but the number of those below $50 million did (figure 5.7).

The data are even more telling when we look at IPOs raising less than $25 million. After increasing to close to 350 IPOs annually from 1993 to 1996, these sub–$25 million IPOs dropped precipitously to under twenty-five by 2000.[3] The sharp decrease began in 2007, just as online brokerage accounts began to proliferate and a full five years ahead of further regulations such as the Sarbanes–Oxley Act of 2002.

Just as a fund size of $100 million to $400 million was the foundation for deep science venture funds, the sub–$50 million IPO was the foundation for the venture capital IPO market and the "anchor leg" of the deep science investing continuum. As the small IPO dissipated from 2001 to 2011, it had a profound impact on the venture capital industry that had traditionally supported deep science investing.

When we review the amount of venture capital raised during the bubble years of 1996 to 2000, we would expect to see far more small IPOs and more IPOs in general between 2001 and 2010. Venture capital raised $243.6 billion between 1996 and 2000, up from $28 billion between 1991 and 1995, and more than the $198.1 billion raised between 2001 and 2008. Yet, rather than all that money flowing into venture capital creating more IPOs, the number of IPOs declined rapidly. After increasing from nearly twenty venture-funded IPOs in 1997 to more than one hundred venture-funded IPOs in 2000, the number of venture-funded IPOs fell to well below thirty per year over the next five years.[4] Clearly, something was happening in the markets to stem small IPOs.

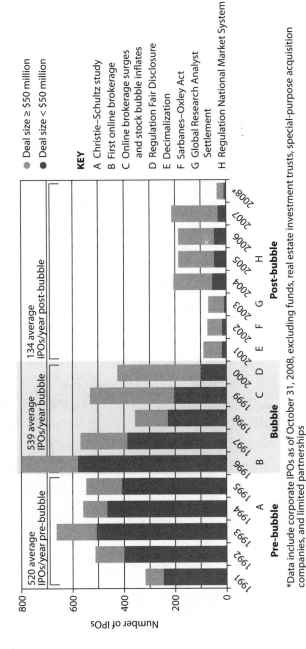

FIGURE 5.7 Number of IPOs, 1991–2008: Deals smaller than $50 million in proceeds versus deals equal to or larger than $50 million in proceeds. *Sources*: Dealogic, Capital Markets Advisory Partners.

*Data include corporate IPOs as of October 31, 2008, excluding funds, real estate investment trusts, special-purpose acquisition companies, and limited partnerships

Additionally, according to the National Venture Capital Association (NVCA), the median age of a venture capital IPO climbed to 8.6 years in 2007, the longest gestation period on record dating back to 1991.

David Weild and Edward Kim argue that it was not the dot-com crash that caused the dearth of small IPOs after 2001. Instead, it was a series of changes that beset the public markets that ultimately caused the small IPO market's collapse. It is worth understanding how the structure of the public market changed if we want to understand if the small IPO, and hence the incentive system for deep science investing, will return at some point in the future.

One significant change came with the emergence of online brokerage accounts in 1996 with Charles Schwab and Co., quickly followed by Datek Online Brokerage Services, E-Trade Financial, Waterhouse Securities, and numerous others. Initial brokerage fees were around $25 per trade, greatly undercutting the whole advice-based brokerage industry, which had charged fees of $250 or more.

Weild and Kim note that "while it is impossible to establish cause and effect, it is reasonable to hypothesize that the dotcom bubble masked an underlying pathology: that the explosive growth in sub-$25-commission-per-trade, self-directed online brokerage accounts brought unprecedented investment into stocks, helped to cause the bubble, and destroyed the very best stock marketing engine the world had ever known."[5] The result was that retail stockbrokers were chased from the no longer sustainable commission-per-trade business of traditional stock brokerage to becoming fee-based financial advisors.

The resulting decrease in traditional stockbrokers had a material effect on the microcapitalization market for public stocks. Stockbrokers, paid $250 or more per trade, had historically scoured the markets for good growth opportunities, often in the microcapitalization market, and brought those ideas to their clients. In effect, they were the stock market business development arm of the microcapitalization companies. These brokers brought demand for stock to these companies, permitting them to grow as public companies.

A second major change came in 2001 with decimalization. Philosophically, the change from trading in fractional increments to decimal increments should have reduced the cost of trading and been a net positive for investors. However, by reducing the increment to $0.01, the incentive for making a market in smaller and riskier stocks disappeared. Trade execution became automated. Market makers, who had previously received $0.25 per share now received $0.01 per share and effectively disappeared from making a market in the smallest stocks. Liquidity dried up.

Unlike larger capitalization companies, micro- and small-capitalization companies required broker-dealers to support liquidity, sales, and equity research, as well as to sustain an active market. One consequence of smaller trading spreads and tick sizes was the tremendous decline in the number of listed companies. Public company listings peaked in 1997 at 8,823 exchange-listed companies and then decreased almost in half to 4,916 listed companies in 2012.[6] That period included fifteen consecutive years of lost listings.

Hedge funds and other hyper-trading institutions became the dominant force in this new market, most likely at the expense of long-term fundamental investors and intermediaries such as liquidity providers. David Weild, Edward Kim, and Lisa Newport note in a paper that market-making that uses capital essential to sustaining quality small companies has disappeared from the now prevalent electronic order books.[7] Firms now pursue strategies as proprietary traders. Market making is becoming extinct.

In their 2011 paper, Weild and Kim conclude,

Generally speaking, economists and regulators have maintained that competition and reduced transaction costs are of great benefit to consumers. This is only true to a point. When it comes to investments, higher front-end or transaction costs and tax structures that penalize speculative (short-term) behavior can act as disincentives to speculative behavior and create incentives for investment (buy-and-hold) behavior that may be essential to avoiding boom-and-bust cycles and maintaining the infrastructure necessary to support a healthy investment culture. As markets become frictionless (i.e., when there is little cost to entering into a transaction), it becomes easier for massive numbers of investors to engage in speculative activity. This occurred first with the introduction of $25 per trade online brokerage commissions in 1996 (which later dropped to below $10 per trade) and decimalization in 2001. Consumers flocked to the markets.[8]

The third change was the introduction of the Regulation Fair Disclosure and the Global Settlement regulations in the United States between 2000 and 2003, which devalued stock research that was critical to microcapitalization stocks. A survey of the literature in a 2007 Harvard Business School study concluded that, by restricting banks from using equity research to support banking, the Global Settlement affected the model used by banks to fund research. This led to a reallocation of research in U.S. equity markets from the banks to brokerage firms, research boutiques, and buy-side firms. Research indicates that analysts at the banks provided less biased and higher-quality research than analysts at these other types of firm.[9]

Sell-side stock research survived only through the commissions it could generate, so the companies covered, and the type of research preferred, changed to meet this need. Quality analyst coverage for the smaller public companies effectively disappeared because it could no longer be supported. These microcapitalization company stories needed to be understood and told, and the Wall Street banking firms, especially some of the smaller ones, were where these stories were told. Regulation Fair Disclosure resulted in institutions halting their payment premium for research. With stockbrokers unable to earn a proper commission, research on the retail side of the business was diminished. Quality sell-side analysts left Wall Street to work at hedge funds. The "dumbing-down" of stock research was in full swing, and smaller companies were left without coverage or with increasingly ineffective coverage.

As a result of regulations that have changed the nature of investment banking, the number of smaller investment banks has also collapsed. The small investment banks had been instrumental in taking companies that raised less than $50 million in their IPOs public, then supporting them with research and trade. Over the past two decades, there has been a stunning decline in the number of investment banks on Wall Street, many of which were well suited to fund science-based companies (table 5.1); this significant decrease has contributed to the diversity breakdown in financial markets. In table 5.1, the names not bolded are the investment banks that have either been acquired or are no longer operating.

For the future of deep science investing, it would be beneficial to have more investment banks participating in the deep science innovation ecosystem. Financial markets function well when they are democratized and when there is a diversity of banks and investors able to shoulder the risks inherent in funding scientific ventures.

The result of this trend in the capital market ecosystem has been a collapse in the support for smaller public companies. Even if smaller companies can go public using one of the larger banks, the resulting marketing and analyst coverage ends quickly. Our experience is that within months of the IPO, the investment banks that took the company public have migrated almost entirely out of a leading market maker position. They effectively abandon the stock.

It is well known that on Wall Street, a very small number of giant funds control the vast majority of trading commissions to the Wall Street banks.[10] These funds have become the best customers not because they are long-term holders of companies but because they trade so frequently. These are high-speed and fast-trading hedge funds. Because such a small number of firms dominate almost all

Table 5.1 Investment Banking System Collapse, 1994

AB Capital and Investment

Advest

AG Edwards and Sons

Allen and Co.

Americorp Securities

Anderson and Strudwick

AT Bred

Auerbach, Pollak and Richardson

Banc of America Securities

Baraban Securities

Barber and Brenson

Baring Securities

Barington Capital

Barron Chase Securities

Beacon Securities

Bear Stearns

Brenner Securities

Chase H&Q

CIBC World Markets

Citigroup Global Markets

Commonwealth Associates

Comprehensive Capital

Craig-Hallum Group

Credit Suisse First Boston

D Biech

Dain Rauscher Wessels

Daiwa Securities America

Dean Witter Reynolds

Deutsche Bank Securities

Deutsche Morgan Grenfell

D. H. Blair

Dickinson

Dillon Gage Securities

Donaldson, Lufkin and Jenrette

Equity Securities Investment

Everen Securities

FAC Equities

FEB Investments

First Asset Management

First Equity Corporation of Florida

First Hanover Securities

First Marathon

Friedman Billings Ramsey

Gilford Securities

GKN Securities

Glasser Capital

Global Capital Securities

Goldman Sachs

Grady and Hatch

Greenway Capital

Hamilton Investments

Hampshire Securities

Hanifen Inshoff

Harriman Group

Harris Nesbitt Gerard

H. J. Meyers

Howe Barnes Investments

IAR Securities Inc.

ING Barings

International Assets Advisory

Investec

Investors Associates

J. Gregory

James Capel

Janney Montgomery Scott

J. C. Bradford

Joseph Stevens

Josephthal

J. P. Morgan Securities

J. W. Charles Securities

Keane Securities

Kennedy Matthews Landis Healy

Kensington Wells

Kidder, Peabody and Co.

Kleinwort Benson Securities

Ladenburg Thalmann

Laidlaw Global Securities

Lam Wagner

Lazard Frères

L. C. Wegard

Legg Mason Wood Walker

Table 5.1 *(continued)*

Lehman Brothers	Paribas Capital Markets	Schroders
L. H. Alton	Parker Hunter	**SG Cowen**
Mabon Securities	Patterson Travis Inc.	Smith Barney
Marleau Lemire Securities	Paulson Investment Co.	Spectrum Securities
Matthews Holinquist	**Piper Jaffray**	Spelman
McDonald and Co.	Principal Financial Securities	**Stephens**
Merrill Lynch	Prudential Securities	Sterling Foster
M. H. Meyerson	RAF Financial	Sterne, Agee and Leach
Miller, Johnson and Kuehn	RAS Securities	Strasbourger Pearson
Montgomery Securities	**Raymond James**	Stratton Oakmont
Morgan Keegan	R. Baron	Summit Investment
Morgan Stanley	Redstone Securities	Texas Capital Securities
Murchison Investment Bankers	Rickel and Associates	Thomas James
NetCity Investments	R. J. Steichen and Co.	Toluca Pacific Securities
NatWest Securities	**Robert W. Baird**	Tucker Anthony
Needham and Co.	Robertson Stephens	**UBS Securities**
Neidiger Tucker Bruner	Robinson-Humphrey	VTR Capital
Nesbitt Burns	Rocky Mountain Securities	Wachovia Capital Markets
Nomura Securities	Rodman and Renshaw	**Wedbush Morgan Securities**
Norcross Securities	Roney Capital Markets	**Wells Fargo Securities**
Oak Ridge Investments	**Roth Capital Partners**	Werbel-Roth Securities
Oppenheimer and Co.	Royce Investment Group	· Wertheim Schroder and Co.
Oscar Gruss and Son Inc.	RwR Securities	Westfield Financial
Pacific Crest Securities	Ryan Lee	Whale Securities
Pacific Growth Equities	Salomon Brothers	**William Blair**
PaineWebber and Co.	Sands Brothers	Yamaichi Securities
Paragon Capital Markets	Schneider Securities	Yoo Desmond Schroeder

Source: David Weild, "The U.S. Need for Venture Exchanges" (presentation, Advisory Committee on Small and Emerging Growth Companies, U.S. Securities and Exchange Commission, Washington D.C., March 4, 2015), slide 12.

the commissions on Wall Street, IPOs are marketed far less broadly than they have been historically. Because of this lack of marketing, when the fast-trading firms sell these new issues, no one is familiar with them and willing to take the other side of the trade and buy the stock. This, along with the other changes discussed, has led to a lack of liquidity in the microcapitalization market, which makes it difficult for small companies to access the additional capital necessary for growth in the public market.

Both the changing structure of public markets and the dearth of small IPOs have been detrimental to deep science investing and economic growth. According to the World Federation of Exchanges, the number of U.S. companies taking their shares off public exchanges has exceeded the number of new listings, resulting in a 36 percent decrease in the ranks of public companies since 2000.[11] David Weild and Edward Kim have reported the same trend, and Jason Voss has extended Weild and Kim's work back to 1975. Voss reports[12] a drop in new equity listings beginning in the late 1990s, with new listings decreasing from nine thousand to five thousand between 1997 and 2012 (figure 5.8). Voss traces this decrease to a loss of ten million jobs over the same period of time, since a lack of listings causes companies to delay their growth and hiring. These are jobs in fast-growing companies that have been decimated by the decrease in IPOs raising smaller amounts of initial capital.

Figure 5.9 shows how each of the regulatory changes maps to the declining number of small IPOs raising less than $50 million between 1991 and 2014.[13]

These changes, together, caused the market structure in the United States to evolve. The current market discourages long-term fundamental investment and encourages short-term trading with no regard to the nature of the company or its business. Volume now flourishes in electronic traded funds and other derivatives, perhaps at the expense of small-capitalization stocks that are less liquid. Fundamentally oriented institutions, which historically sought

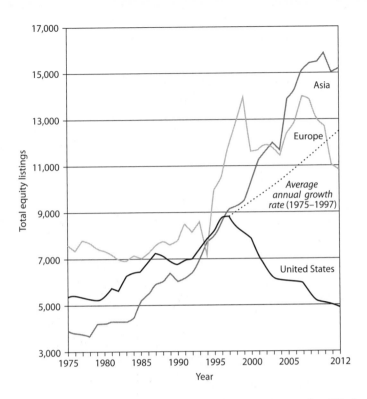

FIGURE 5.8 The decline in American equity listings. *Source*: IssuWorks Inc, 2013–2014.

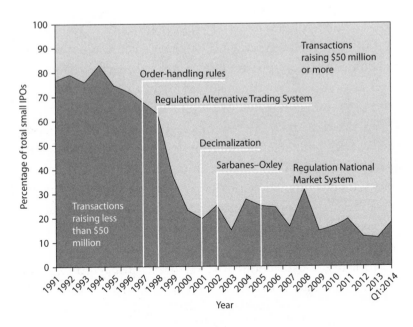

FIGURE 5.9 U.S. regulatory changes drive small-IPO decline. *Source*: IssuWorks, 2013–2014.

growth opportunities in microcapitalization stocks, have abandoned the segment in search of greater liquidity and market scale.[14]

The results of these structural changes to the public market have been catastrophic for deep science investing. As the small IPO—the anchor leg of the relay—disappeared, the entire race for deep science investing began to falter.

One might view the smaller IPO market as risky because of the unseasoned nature of these listings. And while it is true that many of these companies fail, they can also provide the type of growth that has the potential to return many times their initial investment to public shareholders. By getting these companies to the market earlier, with earlier IPOs, these companies can grow and prosper with a larger, more diverse set of investors. The public market, by its very nature, offers a huge diversity of ideas and opportunities for investment. Fidelity, for example, has in excess of $1 trillion of retail money in which a large majority is required to be invested in public companies owing to liquidity requirements and the transparency requirements for public companies.

By enabling earlier IPOs, more of these smaller companies would be able to obtain necessary funding, and the retail investor would be able to participate in IPO returns again. What is gained when companies stay private longer, as in the case of Uber? The answer is that a very, very small set of institutional investment funds reap the vast majority of all returns. Who is winning in the IPO market right now? A few of the largest investment banks, a small set of bankers, and the tiny fraction of institutions that control the vast majority of the commissions on Wall Street. This situation is not aligned with the American democratic ideal of diverse ideas and opportunity. If we go back to relying on early IPOs, we may be able to bring a diversity of opinions and ideas back into investing.

Historically, most venture capital firms raised just one fund and then went out of existence, and so almost two-thirds of the venture capital firms existing in 2000 were gone by 2015.[15] Most of the firms that survived the difficult years of 2000 to 2012 have pivoted quickly to software investments, where the time frame and growth trajectories more closely match the shorter investment time frames of current investors. Later-stage investment funds rose up to take advantage of companies that had no additional access to capital, but even the late-stage funds that have invested in deep science have suffered because the small-IPO market was still not open to them.

Concurrent with this trend, early-stage deep science investments by venture capital firms plummeted. Releases from the *Money-Tree Report* by PricewaterhouseCoopers and the NVCA (based on data from Thomson Reuters) have documented this over the past decade. In 2012, it was reported that despite some fascinating advances in the science of genomics, microRNA, drug delivery, diagnostics, vaccines, and more, the amount of money for first-time company financings in these fields fell by over 50 percent.[16]

Less money is going into the industry, and this is not a one-time quarterly blip but a long-term trend. The venture capital industry has already contracted dramatically, as there were 1,022 active firms during the tech bubble year of 2000, but just 462 active firms left by 2010, according to the NVCA. The NVCA defines "active firms" as those that have invested at least $5 million into companies in the past year. But the number of active firms is still declining, because many firms raised their last funds before the financial crisis of 2008. Most of that pre-2008 money had already been invested in companies, and the firms cannot go back to request that their investors at pensions, endowments, and foundations provide more money, at least until they can generate better returns. With slim odds of taking their portfolio companies public or striking a large acquisition by a big pharma company, the deep science venture capital industry has been withering on the vine for the past four years.[17]

SHORT-TERMISM IN PUBLIC INVESTING

When speaking to domestic investors, it is difficult to inspire conceptual thinking beyond three to six months. Sometimes, it seems that thinking long-term in the United States means looking twelve months out. Contrast this to investors based in the United Kingdom: Invesco Perpetual is a very well-known United Kingdom–based investment firm, and their average holding time for their portfolio is nearly seventeen years! And Invesco Perpetual is not unusual. Woodford and SandAire, both based in the United Kingdom, also invest and hold positions over generations.

Investor time horizons have become an important topic in American financial markets, as there is a growing perception that investors and corporations are becoming myopic and ever more focused on the short term. There has been much discussion in the media in recent years regarding the shortening investor time horizons in U.S. capital markets. As we have discussed, the commercialization process associated with transformative technologies based on deep science is often long, extending over many years and, in some cases, over two decades (see the Nantero case study in appendix 2).

Although there are exceptions, investors in publicly traded securities tend to have shorter time horizons. Michael Dell started a company selling personal computers over the Internet from his college dorm room at the University of Texas, completing a successful IPO years later and then taking his company private in 2013. Dell notes that many shareholders today behave more like renters of equity instead of business owners. His decision to take Dell private reflected a desire to become, as he put it in a note to shareholders, "more flexible and entrepreneurial as a private company, allowing it to do what it does best—to serve customers with a single-minded purpose and drive the innovations that help them achieve their goals."[18]

A growing chorus of technology executives today share Dell's view about the tendency of investors in publicly traded securities to focus on the short term as opposed to the long term. That said, many are reluctant to state their views publicly, given that they are currently running companies with shares listed on the U.S. stock exchanges. As investment strategist Michael Mauboussin notes, "short-termism"—that is, the tendency to make decisions that appear beneficial in the short term at the expense of decisions that have a higher payoff in the long term—is a major issue in the investment business today.[19]

Our personal experience with investors in microcapitalization stock from 2003 through 2007 is that many hedge fund investors who invest in transactions in the microcapitalization market space have holding time horizons of less than one year. In secondary transactions in the public market, we found that, in fact, over two-thirds of the investors in these transactions were no longer investors in the company ninety days after the transaction. When you raise capital to build something that will take many years before value can be realized, there is an extreme mismatch in the maturities of the capital invested and that same capital deployed.

Applying the deep science of information theory to the monetary system in his book *The Scandal of Money*, George Gilder explains the apparent shrinking time horizons of investors and executives in corporate America as a function of a financial system unhinged by antiquated monetary policies and burdensome regulations spiraling downward in a sea of chaos.[20] Chaos in money, says Gilder, stultifies the entire economy. Monetary manipulations of the kind experienced in the United States over the past decade, with interest rates plunging toward zero, shorten the time horizons of the economy. Near-zero interest rates generate what Gilder refers to as "free money," which, in turn, shrinks the horizons of future enterprise. The result is the suppression of entrepreneurial knowledge needed to fuel growth and economic dynamism over time.

Short-termism has caused a great deal of hand-wringing and is said to plague all parties in the investment community, including investment managers, companies, and investors. The typical story, says Mauboussin, is that investors demand short-term results, forcing investment managers to dwell on immediate gains, which ultimately spur investment managers to press companies for quarterly results.

Mauboussin observes that the short-termism in the marketplace today may well reflect the accelerated pace of and exponential growth of technology and the sensation that the world is speeding up. In a world of accelerating technological change, betting on the future feels imprudent if change is rapid. If the sustainability of a company's economic profit is fleeting, there is less reason to place great value on the future.

Figure 5.10 illustrates the acceleration of U.S. technological change and its impact on households. The graph shows how long it took half of the U.S. population to adopt a number of significant new technologies, several of which are general-purpose

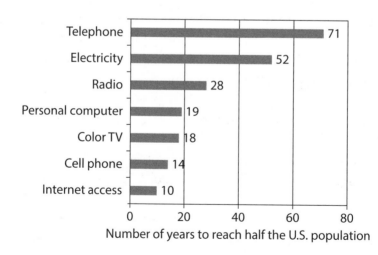

FIGURE 5.10 The diffusion rate of new technologies. *Source*: Adam Thierer and Grant Eskelsen, *Media Metrics: The True State of the Modern Media Marketplace* (Washington, D.C.: Progress and Freedom Foundation, 2008), 18.

technologies. It took just over seventy years for half of the U.S. population to have a telephone and a little more than fifty years for electricity, but only nineteen years to acquire a personal computer and merely a decade for Internet access. As diffusion rates speed up, so does the pace of change in the marketplace. This helps explain the shrinking time horizons observed among economic agents that are contributing to short-termism.

While there appears to be a broad consensus that short-termism exists, much of the evidence for short-termism is based on the perceptions of individuals rather than on the evidence of the stock market itself. What's more, some reduction in investment horizon may be the result of factors relating to the pace of technological change. Nonetheless, as Mauboussin notes, there remains palpable pressure for investment managers, companies, and investors to perform in the short term.

The Stranglehold of Regulation

Regulation is one of the driving forces behind the change in the structure of the public markets. The Sarbanes–Oxley and Dodd–Frank Wall Street Reform and Consumer Protection acts, and additional regulatory changes going all the way back to the negotiated commission decisions of 1975, have compounded on each other to create a very difficult environment for smaller public companies. As venture capital firm Harris & Harris Group founder, Charles Harris, always used to say, "Regulation favors the incumbents." These regulatory changes are part of a larger trend toward increased U.S. regulation highlighted in Charles Murray's 2015 book, *By the People*.

As Murray notes in his book, the landmark Supreme Court decisions from 1937 to 1942 irreversibly increased the range of legislation that Congress could pass and the activities in which the executive branch could engage. The result has been a massive increase in regulation and in the complexity of the legal system. For example, the Sarbanes–Oxley Act of 2002, overhauling rules for financial disclosure by corporations, is 810 pages long. The Patient Protection and Affordable Care Act of 2010 (commonly referred to as Obamacare) is 1,024 pages long. The Dodd–Frank Wall Street Reform and Consumer Protection Act, also of 2010, passed in response to the financial crisis of 2008, is a whopping 2,300 pages long. The U.S. tax code—at some four million words—is about five times the length of the King James Bible and is riddled with ambiguities and special provisions.[21]

Murray points out that the complexity of the law often creates a situation indistinguishable not only from lawlessness but from a kleptocracy. Owners of small businesses now routinely pay lawyers they cannot easily afford in order to pursue a decision from the confusion created by the bureaucracy. Often, these are not unusual decisions, but run-of-the-mill permissions to go ahead with some innocuous business activity that is then made confusing within the bureaucracy. The rules and regulations are so complicated that lawyers are required. The regulation always favors the incumbents in an industry, not the entrepreneurs, and is a tax on economic dynamism.

Meanwhile, the adoption of strict liability in the United States is having a detrimental impact on innovation. "Strict liability" means that a defendant could be forced to pay damages even if no negligence was involved. Strict liability was applied to manufacturers of products in the 1940s and has, as Murray observes, become de facto for services. The expected liability costs of a new product are so high that many improvements, including safety improvements, are *not* brought to market. Murray points out that, from the viewpoint of manufacturers of products and providers of services, being found liable these days in the United States can feel like being struck by lightning. Under the strict-liability doctrine, fault is not even an issue. The U.S. legal system does not need to find that a company did anything wrong to make it pay for an alleged breach. This, says Murray, feels akin to lawlessness.

Another major threat to innovation stemming from U.S. legal developments has to do with the rise of administrative courts and Congress's instigation of regulatory-related lawsuits. This legal development, or administrative law, may be the most pernicious with respect to stifling innovation and economic dynamism. Regulatory agencies in the United States today, notes Murray, have close to total de facto independence. There is no jury inside an administrative court. If a company is prosecuted for violating a regulation issued by the Securities and Exchange Commission, the Environmental Protection Agency, the Occupational Safety and Health Administration, the Department of Health and Human Services, the Department of Energy, or any of the myriad other federal regulatory agencies, the defendant appears before an administrative law judge sitting in an administrative law courtroom. An administrative law judge is selected by the agency whose cases he or she will hear and is an employee of that agency.

Murray notes that most rules of evidence used in normal courts do not apply. The legal burden of proof placed on the lawyer making the case for the regulatory agency is "a preponderance of the evidence," not "clear and convincing evidence," let alone "evidence beyond a reasonable doubt" that one is guilty. If the administrative judge thinks that it is a 51/49 percent call in favor of the regulatory agency that accused the defendant, the defendant is considered guilty. If the defendant considers the administrative law judge's decision adverse, an appeal may be made, but only to another body within the agency. In short, the administrative court is a system riddled with potential for bias. What's more, as Murray observes, the regulatory state in the United States is extralegal in the most straightforward sense of the term. It exists outside the rest of the constitutional legal order.

Path to Change

Throughout this book, we have argued that entrepreneurs backed by venture capital have been a vital part of the deep science innovation ecosystem. In the United States, this partnership has historically generated economic dynamism, productivity growth, and an out-sized payoff to government-sponsored R&D through innovation and the creation of new industries and markets created by this innovation.

As figure 5.11 shows, corporations, the government, and philanthropic organizations have provided funding for research within this deep science ecosystem. Entrepreneurs have capitalized on emerging inventions and intellectual property, driving a dynamic innovation process. Up until 2001 in the United States, there had been a robust "relay race" at play in entrepreneurial

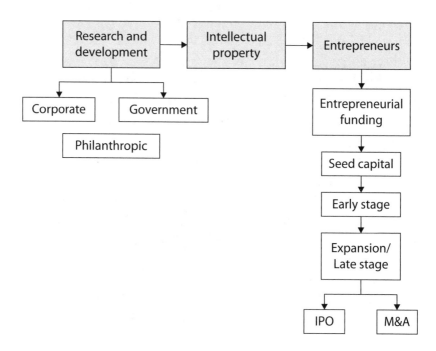

FIGURE 5.11 The deep science innovation ecosystem.

funding, made vibrant through the small-IPO market but then filtering back through expansion-stage, early-stage, and seed capital. Entrepreneurs and venture capital within the American economy have been great beneficiaries, as this innovation ecosystem provides the payoff to R&D.

However, as we have seen in the past few chapters, the multipronged impact of a collapsing small-IPO market, the institutionalization of venture capital, and the rise of software investments that match our shorter-term funding time frames have led to a collapse in deep science investing. Figure 5.12 illustrates the shifting deep science investing landscape. The traditional entrepreneurial funding landscape for deep science innovations that helped fueled economic dynamism and propel Silicon Valley to great heights included a diversity of venture capital deals and a vibrant IPO market. The present entrepreneurial funding environment for deep science innovations lacks the most dynamic elements of traditional funding sources: venture capital and IPOs.

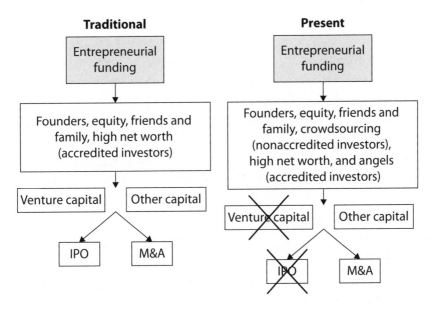

FIGURE 5.12 The shifting deep science innovation funding landscape.

In order to reclaim a dynamic economy, we must find a solution to this current breakdown in the diversity of our innovation investing.

How do we change the current course of deep science investing? Thought leaders like Norman Augustine (former chairman of Lockheed Martin) and Andrew ("Andy") Grove (former CEO of Intel) have discussed the pressing need for deep science innovation today. If the current diversity breakdown continues, the United States is at risk not only of losing the foundations of its technological edge, but of losing the underlying infrastructure needed as the substrate for future technological advancements.

The beauty of the United States is that the oscillations between the cycles we experience in business tend to be shorter than in many places around the world. We recovered from the Great Recession faster than many countries. When something in the investing space swings too far one way, it often corrects itself sooner than in other parts of the world. Part of the reason for this is that Americans are more willing to take risks and more willing to reinvent themselves. Thus, we are starting to see even now that deep science investing is beginning to be appreciated again. Established companies like Google and Facebook, which have been major investors in software and social media, are today investing in deep science and hardware as they look to the future.

One may ask the simple question, "Where will the changes required to foster economic dynamism in the United States come from?" The answer is open to speculation. Our decades of experience in venture capital and financial markets inform us that a number of changes will be required to bring diversity back into the process of funding transformative innovations derived from science. The diversity breakdown in venture capital is just part of a large diversity breakdown in U.S. financial markets that has made it very difficult for entrepreneurs to commercialize scientific products today.

The United States needs more small IPOs, more liquidity in the microcapitalization market, and more smaller investment banks providing research and marketing to smaller public companies. However, to accomplish these things, the United States first needs more investors with longer-term investing horizons. We believe we will see some very large venture funds with permanent capital going after the deep sciences in the years ahead, but the bulk of the funding behind them is in all likelihood going to come from foreign dollars rather than traditional U.S. sources.

A larger number of investors participating in the future deep science innovation ecosystem would attract a greater number of investment banks. Financial markets function well when they are democratized and there is a diversity of banks and investors able to shoulder the risks inherent in funding scientific ventures. It is noteworthy that there have been positive steps in the United States in recent years to democratize the investment process by making it easier for smaller-cap companies to raise money from a greater number of investors.

The SEC's recent decision to run a pilot study of changing the "tick size" for smaller company stocks as of May 2016 is a step in the right direction and probably one of the best ways to begin bringing back a more vibrant public microcapitalization market for deep science. This highly anticipated two-year test program is designed to determine whether trading the stocks of smaller companies in wider tick sizes, or the difference between what traders bid and offer for the shares, will boost interest in the stocks. The move is a shift from more than a decade of requiring trading in pennies. Wider increments, plan advocates say, will allow traders to reap higher rewards, giving them more of an incentive to trade the stocks and lessen volatility. Critics say its goals are aspirational and will accomplish little else than to make trading in certain shares more expensive.

It is also noteworthy that social media has found an application in the realm of fundraising in recent years with crowdfunding (or crowdsourcing) platforms like Kickstarter. These social media–based platforms are changing the way startup capital can be raised.

Crowdfunding has gained traction with entrepreneurs, as it can be a quick and efficient way to raise seed capital from a diverse group of retail investors. While crowdfunding rules are different across countries, they all employ social media to raise capital.

Crowdfunding may take the place of early-stage deep science investors. As of 2016, the crowdfunding industry is on track to account for more funding than venture capital, according to a recent Massolution report.[22] In 2010, there was a relatively small market of early adopters crowdfunding online to the tune of a reported $880 million. In 2014, $16 billion was crowdfunded, and crowd-funding in 2015 rose to $34.4 billion.[23] In comparison, the venture capital industry invests an average of $30 billion each year.

What's more, while crowdfunding has been primarily retail in nature in its early development, more institutional crowdfunding platforms have been emerging in the market (e.g., SeedInvest, CircleUp). The rise of institutional crowdsourcing platforms could become an important future source of funding for startups engaged in developing scientific technologies, especially if a way can be found to keep this fundraising active through the longer cycle of investment needed for deep science technologies. At the very least, it is a positive democratizing event for deep science investing.

The new rules associated with the Jumpstart Our Business Startups (JOBS) Act of 2012 update and expand Regulation A, an existing exemption from registration for smaller issuers of securities. The updated exemption will enable smaller companies to offer and sell up to $50 million of securities in a twelve-month period, subject to eligibility, disclosure, and reporting requirements. Securities and Exchange Commission chair Mary Jo White has noted that it is important for the Commission to continue to look for ways to facilitate venture capital in smaller companies.[24]

With Title IV of the 2012 JOBS Act, the fastest-growing private companies in the United States can conduct IPOs and raise up to $50 million from any investor. Historically, investing in domestic startups and small businesses was reserved for accredited investors.

Now, after three years of anticipation, the ability to invest in pre-IPO companies will be opened up to every citizen. The new SEC legislation is a breath of fresh air and a welcome development for deep science ventures. It may provide further impetus to the crowd-funding platforms that have emerged in recent years. It remains to be seen how entrepreneurs will take advantage of the new rules and how receptive investors will be to funding early-stage companies. Nonetheless, enabling greater diversity in national financial markets will go a long way toward providing the foundation necessary to support the scientific innovation needed to spur economic dynamism over time.

Over the past four to five years, secondary markets for shares of private companies have developed. To date, it is arguable if this secondary market is even a market for anything but a few of the most highly sought-after unicorns. However, if this market can grow further with Nasdaq's purchase of SecondMarket in October 2015, there may be more emphasis on deep science companies in the market.

Because of the asymptotic nature of the value creation for deep science, as discussed earlier, an active secondary, private market for deep science companies could substitute the role historically played by the microcapitalization public markets. Regulation would diminish, permitting these private companies to focus on building long-term value. However, to be successful, a valid market with many buyers and sellers must emerge. To date, we have not witnessed this transformation.

One change we are witnessing in deep science investing is less democratic. We believe it is a logical response to the nature of value creation in deep science investing, though. For the funds large enough to participate meaningfully, we are seeing very large early rounds of capital raised at the company-formation stage. For example, Flagship Ventures–backed Moderna Therapeutics raised over $600 million in an effective first round of financing, and ARCH Venture Partners–backed Juno Therapeutics raised $300 million in the course of a year in its first two rounds of venture financing.[25] Further, Magic Leap has raised over $500 million in early capital.

These large financings early on in a company's development are a response to the asymptotic nature of deep science value accretion. The financing risk is removed for many years, and investors do not have to worry about how value creation will be recognized with each successive capital raise. Additionally, with that much capital raised, these companies can quickly move into the land of unicorns, an opportunity that puts them over the $1 billion valuation, where the public markets work more effectively than in the microcapitalization market. Juno used its large early capital raises to springboard them into the public markets at a multibillion-dollar valuation, where the market for buyers and sellers is still functioning effectively. Now, with the capital necessary to last many years, the company can focus on execution and clinical trials.

Unfortunately, only a small number of funds and participants are large enough to participate in this type of deep science investing. Additionally, a much smaller number of ideas will receive all the capital, meaning that the diversity of ideas will decrease. We do not believe this is a very democratic or efficient way to stimulate deep science innovation, but it is a response to the current difficulties in the market place for those who can afford to invest substantial capital.

As the examples provided demonstrate, solutions are emerging, although none of them is yet developed enough to change the current course of deep science investing. We are, however, optimistic that the tide is changing. Google, Facebook, and other software firms are beginning to turn their attention to deep science. Investors who have fled deep science seem to be slowly moving back into this space. There is a growing recognition today in the United States and around the world of the need for new approaches to fostering the innovation associated with deep science. More and more groups and coalitions are being formed to accelerate innovation in segments such as energy, transportation, machine learning, and space exploration. In late 2015, Bill Gates and a group of technology billionaires, including Jeff Bezos (founder of Amazon), Richard Branson (founder of Virgin Group), Jack Ma (executive chairman of Alibaba Group), and Mark Zuckerberg (founder of Facebook), along with the University of California, announced the formation of the Breakthrough Energy Coalition and an initiative called "Mission Innovation."

The Breakthrough Energy Coalition is based on the principle that technology will solve our global energy issues. It is designed to be a public–private partnership among governments, research institutions, and investors. The coalition has stated that "the existing system of basic research, clean energy investment, regulatory frameworks, and subsidies fails to sufficiently mobilize investment in truly transformative energy solutions for the future."[26] The group aims to invest broadly in clean technology, across electricity generation and storage, transportation, industrial use, agriculture, and energy system efficiency. Twenty countries, including the United States, India, and China, have committed to double their clean energy research investments within the next five years.

It remains to be seen how effective the Breakthrough Energy Coalition will be in fostering the commercialization of clean energy technology in the years ahead. Entrepreneurs will clearly be needed to help drive the commercialization process, and there will undoubtedly be additional investment capital required to push deep science innovation forward. The next five years will be telling for the course of deep science innovation in the United States.

If we are successful in returning deep science investing to its rightful place, the future is bright. There has certainly not been a dearth of new ideas for future computing platforms, new forms of artificial intelligence, new drugs, new methods of prolonging health, new energy solutions, or new ways to feed the planet. Science has continued to move forward rapidly. In the concluding chapter, we will turn toward that potential bright future with the hope that we will be able to return the payoff to government-funded R&D to its original place as the driving force in the economy.

6

Deep Science Venture Investing

When we think about the future, we hope for a future of progress.
—PETER THIEL

REVIEWING THE HISTORY OF deep science investing demonstrates
the importance of recognizing the emergence of a new technology
that will *create entirely new markets or industries*. Historically,
venture capital has been a large contributor to at least six new
industries that it was intimately involved with creating: the microprocessor revolution, the video game industry, the personal computer, biotechnology, telecommunications (hardware for sending
large amounts of data), and, more recently, software and digital
industries enabled by the Internet.

The vast majority of returns in venture capital from the 1960s
through the present have come from the creation and proliferation of these fields. The collaboration between entrepreneurs and
investors has proven to be potent in fostering innovation and promoting economic dynamism. It is safe to say that a large share
of innovation witnessed in the second half of the twentieth century would not have occurred without venture capital–supported
entrepreneurs.

Technological innovation leads to new products, methods, models, and ways of doing business; entrepreneurs and venture capitalists work collaboratively to bring new technologies and products to market. They are the key constituents of the dynamic process associated with the capitalist system that fosters economic growth and prosperity over time. Given the impressive track record of entrepreneurs and venture capital in the latter half of the twentieth century—a track record that many around the world seek to emulate today—there is a much greater appreciation among economists, executives, business leaders, and policymakers of the importance of technological innovation, entrepreneurs, and venture capital in the U.S. economy.

"Disruptive innovation" is a term associated with the work of Harvard professor Clayton Christensen, among others, used to describe innovations that perform the role highlighted by Joseph Schumpeter. While sustaining innovations improve existing products, disruptive innovations are associated with new transformative technologies—technologies that result in new products, methods, models, and ways of doing business. Advances in deep science lie at the foundation of disruptive innovations that, in classic Schumpeterian fashion, have transformed the way people live, work, and play.

Emerging Deep Science Venture Opportunities

The "home-run" game of venture capital investing is unlikely to change any time soon. Such is the nature of venture capital. One of the key questions facing venture capitalists today is where the future home-run opportunities will be. Software investments have been a magnet for attracting venture capital seeking home-run investment opportunities. While venture capitalists may well continue seeking investment opportunities in this segment, there is a growing list of deep science venture investment opportunities that may provide not only home runs, but also serve as a catalyst for future U.S. economic dynamism. Many of these investment opportunities require new technical competencies and new business models, so they fall within the disruptive, radical, and architectural quadrants of the Pisano framework (see chapter 3).

There is a profound evolution underway that is inexorably linked to advances made in deep science and technological innovation over the past four centuries. At the core of this trend is a progression away from "Newtonian" machines toward "Quantum" machines (box 6.1).

The machinery associated with Newton and classical mechanics was a powerful economic catalyst that germinated into the Industrial Revolution. Venture capitalists as we know them today were not around to capitalize on the many opportunities spawned during the Industrial Revolution. That said, many a fortune were made during this time, and the economic effects of the Industrial Revolution on living standards were profound and lasting.

The deep science associated with the Industrial Revolution and the machinery that propelled it contained no independent intelligence, but rather multiplied and leveraged the existing physical capabilities of humans. The advances in deep science associated with Faraday and Maxwell, however, produced the next evolution of machinery: those powered by electricity. The electrification of machinery ushered in a massive wave of Schumpeterian *creative destruction* and laid the foundation for the next evolution of machinery that arose with Planck and the field of quantum

BOX 6.1
The Progression of Deep Science Technology

1600s and 1700s: Newton—mechanical machines
1800s and 1900s: Faraday and Maxwell—electrical machines
2000s: Quantum age—intelligent machines

mechanics. The technologies that lie at the heart of intelligent machinery emerging in the present century have their roots in the deep science of quantum physics.

The Age of Intelligent Machines

The machinery associated with quantum science is of a different character than the machinery associated with the age of Newton or Faraday and Maxwell. The quantum machinery emerging today not only augments the physical capabilities of humans but also our mental abilities. The development of machines that expand the reach of human mental abilities is one that has the potential to unleash a wave of creative destruction that will rival or exceed the first Industrial Revolution. As deep science inventor and futurist Raymond Kurzweil notes, the age of intelligent machines promises to transform production. Education, medicine, aids for those with disabilities, research, the acquisition and distribution of knowledge, communication, wealth generation, and governmental conduct are all likely to be affected.[1]

The potential exists, says Kurzweil of intelligent machines, to begin to solve problems with which humanity has struggled for centuries. These problems include human sensory and physical limitations associated with blindness, deafness, and spinal cord injuries, and new bioengineering techniques that may spawn effective treatments for a wide range of diseases, including genetic disorders. Kurzweil foresees the spectacular 600 percent increase in real per capital wealth over the past century continuing in the twenty-first century as the age of intelligent machines evolves.

Surveying the emerging technological landscape associated with advances in deep science, one cannot help but think, as Kurzweil does, that we have entered a period of economic transformation on par with or larger than the first Industrial Revolution. Today, one frequently hears terms such as "machine learning," "self-driving cars," "drones," the "Industrial Internet," and "additive manu-facturing." Machine learning is the science of getting intelligent machines (i.e., computers) to act without being explicitly pro-grammed. In the past decade, machine learning has spawned self-driving cars, practical speech recognition, and a vastly improved understanding of the human genome.

Deep science researchers see machine learning as progressing toward human-level artificial intelligence (AI). The median view of AI practitioners today is that we are still several decades from achieving human-level AI. Kurzweil is more optimistic and believes the technology will arrive during the next fifteen years—by 2029.

When one hears the terms "AI" and "intelligent machines," one cannot help but think of robotics, which has long captured the fascination of science fiction. The ascent of intelligent machines in the form of robots is accelerating as the twenty-first century progresses. Like the machines that populated the economic land-scape in the wake of the Newtonian revolution in deep science, the first generation of industrial robots did relatively simple, repetitive tasks on production lines where labor was expensive and fault tol-erance was low. Such machines brought precision to automobile production facilities and semiconductor fabrication plants.

The new generation of robotics, with ever-increasing intelli-gence, has the potential to fundamentally change the nature of industrial automation. As the astute global economic strategist and investor Louis-Vincent Gave observes, the current generation of robots can be programmed to undertake complex tasks that previously could be done only by physical labor. These same intel-ligent machines can then be reprogrammed to do different tasks.[2]

Gave notes that in a move reminiscent of the General Motors purchase of the Los Angeles, San Diego, and Baltimore tramways in the 1950s, Amazon spent $775 million in 2012 on Kiva Sys-tems, a supply chain robot maker. Meanwhile, Foxconn in China is experimenting with robotic production lines. Very soon, notes Gave, large-scale robotic adoption and production by firms such as

Amazon and Foxconn will fundamentally change the competitive dynamics of their entire industries. Meanwhile, Google's acquisition of Boston Dynamics in 2013 positions the company as one of the most prominent robotics groups in the country. Eric Schmidt, chairman of Alphabet, Google's parent company, notes that AI and advanced automation make people more productive and smarter.[3]

Quantum science lies at the foundation of some of these transformative deep science technologies, such as robotics and advanced automation. Chief among these are new forms of computing technologies, including quantum computing—a direct byproduct of the development of quantum mechanics in the previous century.

At the forefront of the evolution toward intelligent machines, AI, and advanced automation are computing technologies. Nowhere have advances in deep science been more economically impactful over the past century as in the field of computing. The development of quantum physics since 1900 has fundamentally transformed the economic landscape in a manner as profound as the Newtonian revelation that led to the Industrial Revolution. And it has done so through inventions and innovations associated with computing technology.

As we have noted, Silicon Valley and venture capital have played a prominent role in the quantum revolution in technology. Silicon Valley recently celebrated the fiftieth anniversary of Moore's law. The law, associated with Intel cofounder Gordon Moore, posits a doubling of the power and a halving of the price of computer chips every eighteen to twenty-four months. There is a powerful exponential trend associated with Moore's law—one that continues to astonish analysts and executives who tend to interpolate trends in a linear fashion. The exponential growth associated with quantum technologies is a powerful catalyst for innovation. Analysts estimate that the market value of all companies enabled by the microprocessor and Moore's law is around $13 trillion— equivalent to 75 percent of the current nominal annual output of the U.S. economy—the largest economy in the world.

It is mind-boggling to consider the penetration and economic impact of the microprocessor since 1971. The microprocessor ushered in a new era of integrated electronics that continues into the current century. Microprocessors are embedded in a vast number of products, including computers, electronic devices, cars, trains, planes, and kitchen appliances. Thanks to microprocessors and Moore's law, people carry handheld computers called smartphones, which are more powerful than the biggest computers made in the 1960s. Without microprocessors and the innovations associated with quantum technology, there would be no laptops and no computers powerful enough to chart a genome or design modern medicine's pharmaceuticals. Streaming video, social media, Internet searching, the cloud—none of these activities would have been possible on today's scale without the scientific invention of the microprocessor or Moore's law.

Nearly all of the companies in Silicon Valley today owe their existence to the deep science microprocessor and Moore's law. In the microprocessor, we find the magnum opus of deep science transformative technologies. As former Intel executive Federico Faggin has observed, "The microprocessor is one of the most empowering technologies that mankind has ever produced."[4]

Deep Science Innovation Ahead

As quantum physics penetrates ever deeper into computing technology and machines become more intelligent, we see the potential for a host of transformative scientific innovations with the ability to ignite and propel economic growth and foster future economic dynamism. There are a substantial number of transformative technologies to be developed and commercialized with the assistance of venture investors in the years ahead. Box 6.2 provides a partial list of deep science venture investment opportunities, a list that is likely only to continue expanding.

The opportunities listed in box 6.2 have attracted some interest from venture capitalists and corporations with venture divisions. D-Wave Systems is an emerging deep science firm that is a pioneer in quantum computing. The company has received backing from venture capitalists over the years but is off the radar screen of many venture firms today, given the nature of the technology. It takes a unique breed of venture investor—one with a multidisciplinary background in deep science—to delve into the world of quantum computing and understand the potential of this emerging computing technology in the marketplace. Quantum computing falls into the radical quadrant of Pisano's venture investing framework because it requires new technical skills while leveraging existing business models. The same can be said for other

BOX 6.2
Deep Science Venture Investment Opportunities

- Quantum computers
- Neuromorphic computers
- DNA computers
- Robotics
- AI/machine learning
- Electric vehicles
- Hydrogen fuel cells
- Carbon nanotube memory devices and transistors
- Nanomaterial-enabled photovoltaics
- Next-generation nanomaterial-enabled batteries and energy storage
- Nanomaterial-enabled water filtration and desalination
- Additive manufacturing/3D printing
- Internet of things
- Nano-enabled therapeutics
- Precision health

emerging forms of computing technology, such as neuromorphic and DNA computing.

Meanwhile, there is quite a bit of interest in advanced nanomaterials among deep science venture investors today, given their potential for disruptive innovation and perhaps even architectural and radical innovation. There has been a good deal of commercialization activity with carbon nanotubes (CNTs) being backed by venture capital funds over the past fifteen years. Nantero Inc., is an emerging deep science firm that develops memory devices for computing and other applications using CNTs. The company has received funding from various venture capital firms over the years, but there appears to be limited interest in the company in Silicon Valley at the present time. Nonetheless, Nantero continues to attract interest from other venture investors, both domestically and internationally, and is tracking to commercialize its propriety NRAM technology in the near future.

Additionally, the relatively recent discovery of graphene and 2D advanced nanomaterials has generated a great volume of scientific research and patent activity. Such materials have remarkable properties, including electrical and thermal conductivity, strength (e.g., a single sheet of graphene is one hundred times stronger than steel), and transparency. The unique properties of graphene and other 2D nanomaterials make them attractive for use in innovative commercial applications across a wide spectrum of industries, including energy, electronics, health care, transportation, and water. That said, there is very little interest today among venture capitalists in firms seeking to commercialize graphene and 2D nanomaterial products. Whether the investment climate for such materials will change in the future remains to be seen. As the Nantero case study (appendix 2) demonstrates, the commercialization of advanced materials takes many years and stretches beyond the investment time horizon of the typical general partner–limited partner (GP/LP) venture capital firm structure.

BIOLOGY+, Genomics 2.0, and Precision Medicine

In surveying the deep science investment landscape, there is another significant progression at work in the economy that is attracting the attention of venture capitalists and corporations seeking to innovate. This trend has to do with the use of advanced computing technologies in medicine and health care, beginning with the mapping of the human genome at the turn of the twenty-first century and leading to what are called BIOLOGY+, Genomics 2.0, and precision medicine. As Princeton University scientist Freeman Dyson has observed, "The twentieth century was the century of physics, and the twenty-first century will be the century of biology."[5]

The advent of BIOLOGY+ and Genomics 2.0 opens up the possibility of a world of new innovative applications well suited for commercialization by venture capitalists. BIOLOGY+ engineering, for example, enables microfluidics, which results in low-cost point-of-care medical diagnostics. Having the ability to diagnose in real time during a patient visit allows for a faster response with the correct treatment administered, thereby improving the patient's experience, enhancing provider efficiency, potentially lowering costs, and saving lives with quicker delivery of treatment. BIOLOGY+ material science enables the 3D printing of replacement tissues and organs for use in surgery or drug development. A great deal of progress has been made in moving up the BIOLOGY+ technology learning curve, paving the way to migrate into the application and commercialization phase of development.

Metabolomics and the microbiome represent other exciting frontiers for innovations associated with Genomics 2.0. Metabolomics is a powerful integrative phenotyping technology used for assessing health. Scientists today are building proprietary platforms and informatics systems to foster biomarker discoveries, new types of diagnostics tests, breakthroughs in precision medicine, and robust partnerships in genomics-based health initiatives.

Metabolomics is enabling diagnostics that have the potential to dramatically improve the detection of life-altering diseases such as diabetes and various forms of cancer.

There are tremendous developments taking place in our understanding of the human microbiome. Pharmaceuticals are being developed to treat gastrointestinal issues with beneficial bacteria. Prebiotics and probiotics are going through a renaissance. It turns out that far more diseases than previously known may result from human interaction with the microbiome inside each of us. Our bodies—and the animals and plants we eat—consist of more than just our own cells and our own DNA. Each body is its own ecosystem of bacteria and other microorganisms, all living interdependently. Work in this area reveals that the health of our microbiome has a significant influence on our health.

In the health care and life sciences markets, new tools are being developed permitting us to do more things in vitro (i.e., outside the body) rather than in vivo (i.e., inside the body). For example, as an extension of 3D printing, we will begin to see 3D biology, which includes the printing of organ tissues on a computer chip. Additionally, more medicine is moving to silicon. If one can better replicate the heart on a chip, one can run more toxicity and efficacy studies on those chips. We are in the very early stages of moving some analysis away from laboratory animal studies to those performed on chips.

This research will naturally lead to an increase in silicon research design, which uses advanced computational techniques and machine learning to conduct medical science and research activities with computers rather than with test tubes or animals. Big data, for example, will move into other life sciences applications such as in diagnostics, drug discovery, and the selection of the most effective medical treatments. We have already seen advances in machine learning develop into driverless cars. Increasingly, we will now see big data and algorithmic learning move into life sciences applications as well.

Cell therapy technology—in which cellular material is injected into a patient's usually intact living cells—will continue to progress. A recent example has been the injection of cancer-fighting T-cells through cell-mediated immunity, an area called immunotherapy. Over the coming years, we are likely to see novel materials for cell differentiation, new analysis tools for validating and characterizing cell therapy, and novel manufacturing methods for cost-effective cell production, all of which will make cell therapy a highly viable new tool for disease management.

We characterize Genomics 2.0 as all areas of biological and medical sciences that use new genomic information and approaches as a foundation for their research. But Genomics 2.0 signals a break from the gene-centrism and genetic reductionism of the prior genomic age, which was characterized by a focus on molecular biology at the expense of biochemistry and other phenotypic analyses. A hallmark of this new Genomics 2.0 era is a focus on elucidating the role of the exposome in shaping the process and outcome of gene action. This environment may be the cell, which is the environment of the genes, or it may expand to that of organs or the body.

Genomics 2.0 places an emphasis on complexity, indeterminacy, and gene–environment interactions. Rather than genetics revealing a deep, inner, causal truth, Genomics 2.0 is beginning to conceptualize a complex, relayed, dynamic system of networks of gene–gene interactions, gene–exposome interactions, and highly individualized gene expression and regulation that together produce biological states. Genomics 2.0 is characterized by an emerging scientific consensus that the exposome matters, and a broad commitment to elucidating how gene action and expression are shaped by particular environmental contexts.

The goals of Genomics 2.0 include studying the relationship between genotype, phenotype, proteomics (including structure, expression profile, and interactions), and systems biology (integrating sequence, proteomics, structure, and functional genomics into models of normal and pathological biology; box 6.3).

The ascent of intelligent machines and the development of precision medicine are two deep science–driven trends deserving

---

Box 6.3
The Rise of Genomics 2.0

Extragenic DNA is considered the dark matter of the genome. Only approximately 1 percent of DNA is devoted to protein coding. However, the rest of the genome is transcribed, and it is believed to be involved in a complex level of regulation that we do not yet understand.

Epigenetics is the study of the molecular mechanisms that bring about a heritable or persistent change in gene function without changing the gene sequence.

Proteomics is the large-scale study of proteins, particularly their structures and functions. Proteins are vital parts of living organisms, as they are the main components of the physiological metabolic pathways of cells. The proteome is the entire set of proteins produced or modified by an organism or system. It varies with time and distinct requirements, or stresses, that a cell or organism undergoes.

The exposome is the complement of the genome. Whereas the genome gives rise to a programmed set of molecules in the blood, the exposome is functionally represented by the complementary set of chemicals derived from sources outside genetic control.

Metabolomics is the systematic study of the unique chemical fingerprints that specific cellular processes leave behind. These chemical fingerprints, or metabolites, arise from sources both endogenous to the cell and those that enter the cellular environment. Metabolomics is a biochemical conception of the interaction of the genes within the context of its environment.

With regard to the microbiome, what matters is not necessarily the strains of microbes that make up the microbiome but their genetic resources, and how these cause changes in the way proteins function.

---

attention from venture investors today. A third important trend progression that will also likely profoundly transform the economy as we know it is the convergence of atoms and bits.

The Convergence of Atoms and Bits

Amid all the innovative scientific research and commercialization activity going on at present, there is another emerging technology trend that contains seeds of heightened economic dynamism. This trend is associated with the evolution of 3D printing technology, or "additive manufacturing." At the foundation of 3D printing technology is a convergence of atoms and bits. Additive manufacturing unites advanced nanomaterials and software to form a potent combination that greatly enables technological innovation in ways heretofore unimaginable and uneconomical.

The 3D printing landscape today is reminiscent of the market for personal computers back in the mid-1970s. At that time, there were passionate young innovators with a deep scientific background like Apple cofounders Steve Jobs and Steve Wozniak tinkering with the machines in their garages. To the established computer manufacturers of the time, the personal computer looked more like a child's toy than a business and productivity tool. Some may recall Digital Equipment Corporation president, chairman, and founder Ken Olsen stating in 1977, "There is no reason for any individual to have a computer in his home."[6]

The advent of 3D printing and additive manufacturing is in the early stage of its evolution. As Chris Anderson notes in his book *Makers: The New Industrial Revolution*, the history of the past two decades online is one of an extraordinary explosion of innovation and entrepreneurship. It is now time, says Anderson, to apply this to the real world, with far greater consequences.[7] The emergence of 3D printing is ushering in a new era of innovation and entrepreneurship that has the potential to shift the technological and venture investing landscape away from the present one where software has been "eating the world." Patent activity associated with additive manufacturing is on the rise and is tracing out the familiar exponential growth curve associated with transformative technologies (figure 6.1).

With 3D printing, one envisions the potential for a Cambrian-esque explosion of innovative designs and products akin to the proliferation of various types of machinery that populated the eighteenth century and fundamentally transformed the economic landscape. In the coming era of 3D-printed products, customers will be able to choose from endless combinations of shapes, sizes, and colors. Whereas the first Industrial Revolution was about mass production, additive manufacturing enables mass customization. In sum, the convergence of atoms and bits in the form of additive manufacturing is approaching an inflection point whereby it will become mainstream with game-changing implications for the way we live, work, and play.

It remains to be seen what deep scientific advances lie ahead and what technologies these advances will spawn. There is a great deal of transformative technological innovation already in progress—certainly enough to occupy the most experienced

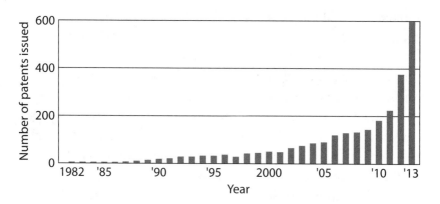

FIGURE 6.1 Additive manufacturing patents issued worldwide, 1982–2013. *Source*: Richard D'Aveni.

science venture investors in the coming years. Where we go from there is anybody's educated guess.

While software investments have been attracting the attention of venture capitalists in recent years, the foregoing discussion demonstrates that there is a host of emerging deep science happening that could be more impactful if it were translated into a payoff by venture capital. Underlying these opportunities are some major trends related to the evolution of deep science and technological change that powerfully drive innovation and foster economic dynamism. The move toward intelligent machines and advanced forms of computing is well underway; attracting more venture capitalists into this arena of deep science venture investing could accelerate its momentum.

The same can be said of two other major trends: precision health and additive manufacturing. Clearly, software will continue playing an integral role in these three trends, and social media may well play an important role as well. Transformative innovations that fall within the various quadrants of the Pisano framework carry within them surprise and thus possess a high degree of entropy from an information theory perspective. It is the surprising or high-entropy nature of transformative deep science innovations that is a catalyst for economic dynamism.

The revolution in science that began in the early part of the twentieth century with the development of quantum mechanics has yet to be completed. The age of quantum computing has dawned but is only in a very early stage of development, as the D-Wave case study (appendix 1) makes clear. The age of nanomaterials, too, is in an early stage of development. Silicon has served an admirable role in the development of quantum technologies, but scientists question its ability to scale below the seven-nanometer level of today's microprocessors. Carbon nanomaterials, whether CNTs or graphene, possess remarkable properties superior to silicon for applications in computing and electronics, not to mention energy, transportation, and water. The Nantero case study (appendix 2) illustrates the hard work, diligence, time, and capital it takes to commercialize innovative CNT memory devices such as the NRAM. Silicon Valley may well be transformed into "Carbon Valley" in due time and usher in a wave of innovation and economic dynamism that rivals or surpasses what has transpired over the past sixty years.

The issue confronting us today is not a shortage of transformative deep science venture investment opportunities, but rather how the technologies associated with deep science will get funded and commercialized in the months and years ahead. As we have noted, software has been "eating the world" in recent years, and venture capital has increasingly migrated away from companies engaged in commercializing transformative technologies based on deep science. In the next chapter, we will discuss the issues surrounding the funding of companies pioneering transformative technologies based on deep science and propose some solutions to spur such funding with the longer-term target of fostering increased economic dynamism in the United States and in economies worldwide.

7

Our Choice Ahead

MEDIA HEADLINES PROCLAIM VENTURE investing is attracting billions of dollars of new capital after years of languishing since the dotcom bust. Fueling the revival of venture investing today is a resurgence of investment activity in Silicon Valley, driven by the success of enterprises like Facebook, Twitter, and Uber. The ascent of software investments over the past decade is transforming the venture capital landscape with important economic repercussions.

The heightened pace of venture investing activity in the Valley has observers wondering whether there is another dotcom bust in the making, but significant changes in the U.S. capital market since 2000 make another such bust unlikely. Our primary concern is not the potential for another meltdown, but rather the concentration of venture capital in the software investment segment. The attraction of software investments is siphoning capital away from transformative deep science technologies and the entrepreneurial companies that commercialize them.

As a sign of the times, IPO activity associated with deep science came to a halt in the United States during the first eight months of

2016. During this time, only one venture-backed technology IPO was completed.[1] This sad state of affairs, we believe, is symptomatic of everything we have discussed in previous chapters. Commenting on the current environment, George Gilder notes that the big Wall Street investment banks have horizons that are far too short to foster entrepreneurial wealth and growth. These institutions, says Gilder, "insidiously thrive today by serving government rather than entrepreneurs. Government policy now favors the short-term arbitrage and rapid trading of the big banks over the long-term commitments that foster employment growth, leaving us with the predatory zero-sum economy that destroys the jobs and depletes the incomes of the middle class."[2]

Although the big banks are catering to government generally, Wall Street's actions are counter to fostering a payoff to deep science R&D. As a case in point, consider the billions of dollars the U.S. government has so far channeled into nanotechnology-related R&D during the twenty-first century. From time to time in our roles of venture capitalists funding entrepreneurs seeking to commercialize transformative nanotechnologies, we get asked by various U.S. government officials why there has not been a bigger payback from the over $20 billion of R&D investment associated with the National Nanotechnology Initiative (NNI) over the past fifteen years.

For those unfamiliar, the NNI is federal government program for the development of nanotechnology. Many federal government agencies are involved with the NNI, including the National Institutes of Health, the National Science Foundation, the Department of Energy, the Department of Defense, and the National Institute of Standards and Technology. The program has received bipartisan support since its inception at the beginning of this century. The mission of the NNI is to advance a world-class nanotechnology R&D program, foster the transformation of new nanotechnology into products for commercial and public benefit, develop a supporting infrastructure of educated and highly skilled workers and tools to advance nanotechnology, and support the responsible development of nanotechnology.

The intention of the NNI program is ambitious and admirable, but, as highlighted in this book, the program was established and developed during a period when the link between government-funded deep science R&D and the payoff to deep science R&D has been severely weakened by both the diversity breakdown in venture capital (noted in chapter 4) and structural shifts in U.S. capital markets (discussed in chapter 5). The lack of deep science venture funding and capital market structural shifts have worked to significantly constrain the achievement of one of the main goals of the NNI: to foster the transformation of new nanotechnology into products for commercial and public benefit. These factors have inhibited the potential to generate a meaningful payoff to the billions and billions of (taxpayer) dollars used to fund nanotechnology R&D in the United States in this century.

As we have argued throughout this book, the effectiveness of federal investments in R&D can be greatly improved if there are systems in place to capture a payoff to R&D. In our view, the NNI's mission of fostering the commercialization of transformative nanotechnologies could be more successful in an environment where there are an increased focus by venture capital on deep science investing and where U.S. capital markets are shaped to be more hospitable to capital access by micro- and small-capitalization companies. As discussed in chapter 5, these factors are probably inextricably linked, meaning that the latter could likely spur the former and provide a positive feedback loop for deep science investing and future productivity.

We are at a critical juncture for American venture investing. The stunning economic successes associated with new markets and industries founded by digital technologies have fostered a second wave of venture investing in software and social media. This wave has led to a resurgence of Silicon Valley and venture capital but has also cut back its diversity, draining resources away from deep science investment opportunities.

Historically, venture capital has played an important role in the payoff to R&D. While R&D still occurs and has grown over time alongside the U.S. economy, the investment to generate the payoff to that R&D is not as robust as it was during the period following World War II through 2000. This trend carries with it ominous implications for American innovation and economic prosperity.

Deep science R&D and technological development have been the predominate sources of invention and innovation underpinning economic dynamism. The term "economic dynamism" has been used in this book to represent the underlying forces in the economy that drive growth in employment, income, output, productivity, and wealth creation. "Business dynamism," another term used in this book, is the process by which firms continually are born, fail, expand, and contract, as jobs are created, others are destroyed, and others still are turned over. Transformative scientific technologies were instrumental in fostering economic growth and lifting living standards to ever-greater heights from the end of the Second World War through the turn of the twenty-first century. To stifle such innovation is to dampen and undermine the economic forces that create jobs, income, and wealth and improve living standards.

Creative Disruption

Advances in deep science have inspired transformative technologies for the past four hundred years. The Newtonian scientific revolution and the innovations that followed are without peer in the annals of economic history—and have dramatically altered the technological and business landscape in profound and often unexpected ways.

The notion that scientific and technical knowledge is vital to American living standards is enshrined in the Constitution, which explicitly gives Congress the power to "promote the progress of science and useful arts" by granting inventors patents. Government officials around the world today recognize scientific knowledge as a key ingredient in long-term economic vitality. We have seen how the Newtonian revolution inspired a new age of machinery that gave birth to the Industrial Revolution. The development of classical physics was transformative, shifting our view of the cosmos and inspiring new technological inventions and ways of doing business. Advances in deep science in the seventeenth century unleashed an economic dynamic never before seen. The proliferation of new scientific technologies led to a revolution in industry that profoundly altered the economic landscape in the form of new businesses, jobs, products, and related services, which, in turn, fundamentally changed the way people worked, lived, and played.

Many technologies spawned over the decades are taken for granted today. Newtonian machinery and Maxwellian electricity are woven so deeply into the fabric of our modern economy that many hardly notice them. Such is the transformative power of deep science technologies that they often disappear into the background and become taken for granted. More recently, this has been the case for nanotechnology as well. One can only imagine the astonishment of an average citizen of the fourteenth or fifteenth century discovering that one could flick a small switch to illuminate a large room or power a piece of electrical industrial machinery.

As economist Scott Grannis notes in a blog post, the daily life of an American billionaire today is not much different from that of his middle-class neighbor.[3] Both have ready access to great food, a clean abode, cheap travel, cheap entertainment, great health care, and instant communications with just about anyone anywhere in the world. Because the U.S. economy can produce so much, so efficiently, says Grannis, the fruits of modern life (e.g., iPhones, air travel, clean water) are available to virtually everyone. This becomes all the more evident when visiting centuries-old castles and estates today from eras gone by where it took an army of servants to provide a standard of living for the wealthiest of land-owners that would be considered below poverty level today. No electric light bulbs. No refrigerators. No HVAC. No TV.

Only a half-century ago, an ambitious scientist with an entre-preneurial spirit named Gordon Moore saw that the number of transistors on a silicon chip could double every year or two, thus providing a powerful impetus to emerging digital computing technology. There are billions of computing devices on the planet today driven by Moore's law that are used by consumers, busi-nesses, and governments. Today, for the cost of an hour of work on the average wage, a typical consumer can buy about a trillion times as much computing power as was available when Gordon Moore wrote his now famous article.[4] Moore's law has had a colossal effect on global living standards over the past fifty years and has few rivals in fostering economic dynamism.

Technological change, driven by scientific advances, has acceler-ated over the past century. The quantum revolution in science is ush-ering in an age of intelligent machinery: powerful computers and devices that promise to transform the world economy. Many nations' priorities today—agriculture, water, energy, health, communications, and transportation—can be addressed through scientific innovation.

Deep science is revolutionary: Newtonian mechanics, Max-well's theory of electromagnetism, Einstein's special and general relativity theory, quantum mechanics, information theory, the central dogma of molecular biology, and the sciences of com-plexity. Deep science fundamentally changes the way we view nature and the cosmos. It inspires new research and investigative methods, which lead to new inventions and innovations. Some are massively transformative and referred to in the economics lit-erature as general-purpose technologies (GPTs). Science-derived GPTs include the steam engine, the railroad system, electricity, the microprocessor, and the Internet.

General-purpose technologies unleash a powerful economic dynamic that pulses through an economy for years. Economists refer to this dynamic process as "creative destruction," a term first used by economist Joseph Schumpeter. Often, deep scientific advances and their technological innovations are feared, as they may disrupt the normal flow of business, jobs, and livelihoods. We see this fear manifesting today in discussions about artificial intelligence and the effect it may have on future employment and society.

Such fears, while understandable, fail to appreciate the vital role scientific technologies play in fostering economic dynamism and promoting prosperity by improving living standards. There was a massive wave of scientifically driven technological innovation in the United States during the twentieth century. During that time, U.S. employment expanded from fewer than 25 million jobs to over 125 million—a more-than-fivefold increase (figure 7.1).

The remarkable expansion of U.S. employment experienced during the last century occurred alongside tremendous technological innovations. While such gains are impressive, they are not as surprising in light of the connection between economic dynamism and scientific innovation. Such appreciable expansion in

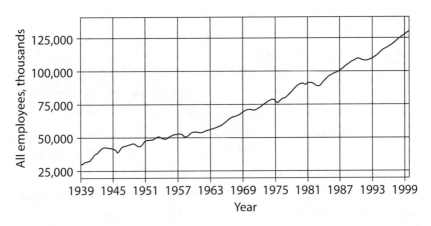

FIGURE 7.1 U.S. nonfarm payroll employment, 1939–1999. *Source*: U.S. Bureau of Labor Statistics.

employment is, in fact, par for the course in an economy driven by advances in deep science. The dynamism linked to deep science accomplishments fuels prosperity.

While these scientific advances have contributed to the quantity of jobs available overall, they have also created different kinds of jobs. The nature of what one does shifts, and the changes in the structure of employment reflect the nature of the technological changes wrought by deep science. This phenomenon is evident in the U.S. economy's experience during the twentieth century. Economist W. Michael Cox notes that the labor market is in a perpetual state of what he calls "churn," a metaphor taken from the way water dynamically swirls and eddies, every movement the result of the same inexorable force.[5] Whereas "downsizing" focuses solely on the discomfiting side of economic change, the churning image captures the whole process: the good along with the bad, the jobs that companies create as well as the ones they let go.

The churn, notes Cox, is not a new phenomenon. Throughout history, each generation of jobs has given way to the next. Over the millennia, as technology progressed, the nature of work changed. Whereas farming and agriculture were a major source of employment in the United States in 1900, today's labor market comprises a diverse array of jobs that did not exist back then. In the early years of the previous century, jobs were concentrated in companies engaged in metals, oil refining, meatpacking, and basic machinery. While these segments were the technology leaders of their day, over the course of the century, jobs shifted to those in consumer goods, technology, retailing, finance, and services.

Cox notes that the churn is not just a matter of creating more jobs, but better jobs. As the U.S. economy has evolved, recycling in the labor market has tended to benefit workers despite the occasional sting of the pink slip, says Cox. On balance, paychecks have grown fatter and workweeks have become shorter. The backbreaking toil of farms and sweatshops has given way to the comfort of air-conditioned offices. The churn reflects the evolution of technology and is propelled by better ways of providing what consumers want as living standards shift over time. Changes in the nature of technology, spurred by advances in deep science, transmit powerful impulses through the economy that churn the labor market.

Job market churn is, in fact, essential to the economic process that facilitates the expansion of employment, income, wealth, and productivity. Labor market flexibility is important for this very reason. The successes Silicon Valley witnessed over the past half century are related to flexibility of the California labor market, which allows resources to flow into more productive areas. In 2015, the state of California enjoyed record levels of employment, with nonfarm payrolls exceeding sixteen million for the first time in the state's history, due in part to a surge in technology jobs.

The complicated dynamics of technological evolution reverberate through an economic system—not over months or quarters, but years. As technology advances, so do the businesses creating new jobs. Absent the dynamism of labor markets, businesses would stagnate over time and so too would the general economy. The natural fluctuations associated with technological change continually revolutionize the structure of the economy from within—in creating, the new displaces the old.

The U.S. economy in the twentieth century showed that the Schumpeterian creative destruction associated with scientific innovations is not a zero-sum game at all. Deep science technologies greatly increase employment opportunities through the expansion of new businesses. Yes, there is churn in the job market, but that churn fosters the reallocation of resources that leads to more jobs, greater income and wealth, and higher living standards. Trillions of dollars of income and wealth flowed from the creative destruction associated with scientific advance and technological change. Significantly, real per capital GDP rose seven-fold in the United States in the twentieth century, from around $6,000 in 1900 to over $43,000 in 1999.

In *Creative Destruction,* Richard Foster and Sarah Kaplan note that all institutions benefit from the refreshing processes that produce churn in the labor market and heighten living standards over time.[6] Failure to provide avenues for continual change—to create new options and rid the system of old processes—eventually causes organizational failure. Foster and Kaplan note that if the forces of creative destruction are suppressed for long periods, the ruptures can destroy institutions and individuals with astonishing speed, as political and military revolutions have taught us consistently.

Foster and Kaplan point out that the benefit of legitimate capital markets is both the coordination of the wishes and capabilities of millions and the peaceful processes that set an appropriate pace and scale for change. Without an active market, entrepreneurship can be suppressed, sometimes for decades. The authors note that the importance of what Schumpeter observed cannot be overstated and that it would be wise for policymakers to take note that "it is not [price and output] competition which counts, but the competition from the new commodity, the new technology, the new source of supply, the new type of organization" (295).

It's All About Jobs

Entrepreneur and businessman Michael Dell, in his role as global advocate for entrepreneurship with the United Nations (UN), penned an article in 2015 titled "Wanted: 600 Million New Jobs." Dell notes that it is critical that entrepreneurs and the jobs they create become part of the UN's Sustainable Development Goals, the world's "to-do list," as he put it, for the next fifteen years. It is critical, says Dell, "because we need six hundred million jobs to employ a rapidly growing global work force." Contrary to popular belief, the new jobs will not come from large corporations but from the activity of entrepreneurs and their fast-growing startups.

Dell notes that entrepreneurs create over three-quarters of all new jobs in the world. It simply will not be possible to create six hundred million new jobs without a healthy pipeline of startups and thriving small businesses. Dell states that it is important to remove the barriers that prevent entrepreneurs from turning their ideas and breakthrough innovations into thriving businesses. The benefits of the work of entrepreneurs are enormous, says Dell, with the biggest benefit of all being hope, "because, wherever you are in the world, jobs and economic opportunity bring hope for a better future."[7]

Meanwhile, American thought leaders are recognizing how critical it is to support science education in curricula nationwide. Prior to his death in 2016, legendary venture capitalist Larry Bock had been especially proactive on this front over the past several years. Bock pointed out that American students are not entering the science and engineering fields.[8] Furthermore, said Bock, we are not retaining the people we are educating from abroad in those fields, owing to visa issues. Additionally, science and engineering opportunities are currently greater abroad than in the United States. The confluence of these trends has led to a perfect storm in which Americans are not entering these vital fields. Bock believed that if this trend is not reversed within one generation, the United States will have primarily outsourced innovation.

Bock was founder and executive director of the USA Science and Engineering Festival, the nation's largest annual celebration of science and engineering. The mission of the festival is to stimulate and sustain the interest of the nation's youth in science, technology, engineering, and math (fields collectively referred to as "STEM") by producing the most compelling, exciting, and educational festival in the world. The festival serves as an open forum to showcase all facets of STEM.[9] The festival attracts tens of thousands of people each year and is serving an important function at a critical juncture in the evolution of the American economy.

The National Science Foundation was founded in 1950 to assist with promoting the long-term strength of the American scientific community. During the foundation's early years, the workforce consisted of scientists and engineers engaged in R&D in government, academic, and industry laboratories. Over the ensuing six decades, policymakers, scholars, and employers came to recognize that STEM knowledge and skills are critical to the American workforce. It is also widely acknowledged that a broad range of STEM-capable workers contributes to innovation and economic dynamism and that the STEM-related workforce is a vital part of the U.S. economy. It is noteworthy that the STEM workforce enjoys greater employment opportunities than other members of the U.S. labor force, as indicated by lower rates of unemployment in the STEM workforce. Further, STEM degree holders at all educational levels experience a wage premium compared with the general workforce.

The STEM workforce fields play a direct and vital role in driving economic growth, and this role is only becoming more important. The federal government invests over $4 billion annually in STEM education and training. Over twenty-six million U.S. jobs, some 20 percent of all jobs available, require advanced knowledge in any one STEM field. Half of all STEM jobs are available to workers without a four-year college degree, and these jobs pay $53,000 on average—a wage 10 percent higher than non-STEM jobs with similar educational requirements. Half of all STEM jobs are in the manufacturing, health care, and construction industries.[10]

Technological innovation that spurs economic dynamism requires the expertise of specialists with knowledge in the STEM fields. The STEM workforce is integral to the science and technology enterprise, and its role has grown significantly since the Industrial Revolution amid the scientific advances of the twentieth and twenty-first centuries (figure 7.2).

The U.S. federal government's explicit commitment to fund the STEM labor supply and to promote research can be traced

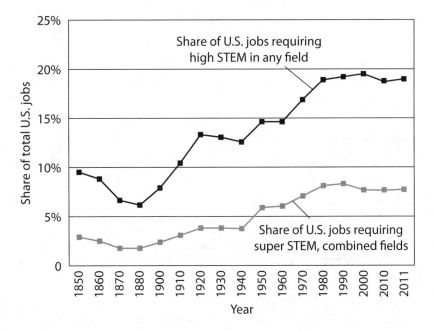

FIGURE 7.2 Share of U.S. jobs requiring a high level of knowledge in a specific STEM field ("high STEM") or a high level of overall STEM knowledge ("super STEM"), 1850–2011.

The Brookings Institution study defines STEM jobs in two ways, the second more restrictive than the first: (1) High STEM in any one field: The occupation must have a knowledge score of at least 1.5 standard deviations above the mean in at least one STEM field; these occupations are referred to as high STEM throughout the report. (2) Super STEM or high STEM across fields: The occupation's combined STEM score—the sum of the scores from each field—must be at least 1.5 standard deviations above the mean score; the report refers to these occupations as super STEM.

Source: Jonathan Rothwell, "The Hidden STEM Economy," *Brookings*, June 10, 2013.

to Vannevar Bush, who helped establish the National Science Foundation. Since that time, reports from the foundation have emphasized the need for STEM education. STEM receives strong support from both sides of the political aisle. In 2006, President George W. Bush launched the American Competitiveness Initiative to improve STEM education and increase the supply of working scientists. More recently, President Obama created the "Educate to Innovate" campaign to boost STEM education and signed into law a reauthorization of the Bush-era America COMPETES Act, which embodies many of the same goals as the Bush administration's STEM priorities.

The National Science Board, the policymaking body of the National Science Foundation and an adviser to the president and Congress, examined recent STEM workforce studies and debates, consulted numerous experts, and explored data in the foundation's 2014 *Science and Engineering Indicators* report to develop insights that could facilitate more constructive discussions about the STEM workforce and inform decision makers. Among the primary insights to emerge from the board's analysis is that the STEM workforce is extensive and critical to innovation and U.S. competitiveness.[11] The important role of the American STEM workforce highlights the vital role of academia in facilitating the education and training required for such jobs.

Labor market trends in the U.S. economy have been anything but robust in recent years. The private sector has created some thirteen million jobs in the past five years, but the net gain from the prior economic expansion's peak has been only about four million. That means a net gain of only about 0.5 percent per year. In the 1980s, when entrepreneurship flourished in the United States, net job gains were around 2 percent per year. The more robust net job gains experienced in the 1980s would have supplied nearly fifteen million more jobs today, more than all the jobs created during the past five years.

A Tide that Lifts All Boats

In light of the trends highlighted here, it is not surprising to see more thought leaders and policymakers in the United States show concern for issues with funding scientific startup enterprises—endeavors that have in the past fueled employment, wealth, and productivity growth. Intel Capital president Arvind Sodhani observed recently that, while software investments receive the bulk of attention today, we are going to need hardware startups if many of the disruptions and innovations, such as driverless cars and drones, we are seeing today are to fulfill their promise.[12] Sodhani notes that there is a lot of capital being invested trying to find the next Facebook but that it is well known that there will likely be only a half-dozen Facebooks in the next decade or so. Sodhani believes such pursuits are harmful.

Amid the trend of software "eating the world" today, there is a chorus of thought leaders both inside and outside Silicon Valley concerned about the fact that new digital technologies are doing little to benefit a large share of American workers. An open letter from a group of Silicon Valley venture capitalists and academics released in spring 2015 highlights the declining economic dynamism in the U.S. economy over the past two decades. The letter points out that the major technological advances of the past—like the electrification of industry—spawned sustained job and income growth. This time around, however, the evidence is causing people to reflect and wonder whether things are different.[13] Those concerned believe there is a great deal that can be done to improve everyone's prospects during a time of transformation driven by digital technologies. They proposed a three-pronged effort:

1. A set of basic public policy changes in the areas of education, infrastructure, entrepreneurship trade, immigration, and research.
2. The development of new organizational models and approaches that not only enhance productivity and generate wealth but also create broad-based opportunity. The goal, says the group, should be inclusive prosperity.
3. More and better research on the economic and social implications of the digital revolution in technology and

increased efforts to develop long-term solutions that go beyond current thinking.

As history has shown, deep science ventures are a tide that lifts all boats. We believe that the time is now to find ways to bring diversity back into investing to support early-stage deep science enterprises. Our future prosperity in the global market depends on it. The ecosystem of American scientific innovation is diverse, including entrepreneurs, venture capitalists, and corporations traditionally playing key roles. The migration of venture capital away from deep science investments and the increasing concentration in software investments are among the most disconcerting trends today, in our view. Because of the move to shorter investment time horizons, software investments have become extremely compelling relative to deep science investments in recent years. The success of software investments has done much to revive the venture capital business in the United States.

That said, there are signs that the diversity breakdown in venture capital financing is moderating. Software investments have become less compelling amid rising startup valuations coupled with the relatively high cost of living in the Valley. Diversity breakdowns, when they occur, often signal turning points. However, it is not yet evident what segments venture capital will flow into following this breakdown. In chapter 6, we highlighted a host of emerging scientific technologies deserving attention from venture investors, but the presence of such opportunities does not mean they will attract the attention of venture capitalists.

While venture capitalists in Silicon Valley and elsewhere continue to shun deep science ventures in favor of software investments, it is encouraging to see established technology companies like Google and social media companies like Facebook stepping up investments in deep science; for instance, in the life sciences, robotics, and machine learning. It is also encouraging to see startup venture capital funds moving away from software investments toward companies focused on longer-term issues. Lux Capital raised $350 million for a new venture fund focused on deep science–based early-stage investments. Two early investors in Tesla Motors teamed up in the summer of 2015 and raised $400 million for such investments, stating that the timing was propitious for such funds.[14]

The partners of the new venture funds noted they are following in the path of deep science venture investor Elon Musk, who has proven that it possible to make money while also improving society. A scientist or entrepreneur aspiring to cure cancer—against all the obstacles such an endeavor involves—is almost certainly not driven by pure financial gain. As veteran Silicon Valley venture capitalist Kevin Fong says, "At a certain point, it's not about the money anymore. Every engineer wants their product to make a difference. You want your work to be recognized for what it does for people."[15] As venture investors Victor Hwang and Greg Horowitt have noted, we are all beneficiaries of the love that has motivated countless scientists, engineers, and entrepreneurs over the years.

Deep science and innovation lie at the core of much of what entrepreneur Elon Musk is doing today. It remains to be seen whether the investment activity spawned by his successes is the beginning of a new trend that will lure venture capitalists back to deep science in a major way. Such a migration would certainly be timely and welcomed after years of neglecting the type of deep science investing so vital to economic dynamism and long-term prosperity.

Venture Investing Scenarios

The future of venture investing has major implications for society. Venture capital has played a key role in fostering the commercialization of deep science and has been a critical catalyst in promoting economic dynamism, which creates jobs, income, and wealth that increases living standards. Venture capitalists have played an instrumental role in helping to realize Vannevar Bush's post–World War II vision of deep science in the U.S. economy. Venture capital, though a small fraction of total investable capital, has had an outsized impact on the creation and success of new products and industries founded on deep science. As we have seen, the ecosystem that drives innovation is diverse, but it is difficult to overemphasize venture capital's position within the deep science innovation ecosystem.

Peter Drucker once observed that every organization needs one core competency: innovation. Apple cofounder Steve Jobs noted that innovation is what distinguishes a leader from a follower in business. The paucity of deep science venture deals today threatens to disrupt the innovation process that is instrumental to every organization and fundamental to economic prosperity. In summarizing the analysis and discussion in this book, we envision two scenarios that may play out in the future.

Scenario 1: The Pessimistic Scenario. In this scenario, U.S. venture capital continues to concentrate in software investments. The diversity breakdown process (discussed in chapter 4) continues to unfold. Silicon Valley stops investing in silicon and shuns deep science deals. As a result, scientific innovation and commercialization increasingly migrate offshore, where there is greater access to capital and a willingness to patiently fund deep science enterprises. American innovation stagnates and competition intensifies overseas, with concomitant negative effects on U.S. economic dynamism. This pessimistic scenario is one that is anathema to the American entrepreneurial spirit and the past history of Silicon Valley. The likelihood of this scenario playing out in the future cannot be minimized at this juncture, for this is the path that venture capital and the U.S. economy have been treading over the past decade.

Scenario 2: Turnaround Scenario. This scenario is more positive and more in line with the American entrepreneurial spirit and the past success of Silicon Valley. In this scenario, venture capitalists increasingly turn their attention to deep science investments—robotics, artificial intelligence and machine learning, precision health and medicine, and quantum computing, thus ending the diversity breakdown in American venture capital investing. The renewed attention on deep science deals will reflect the growing investment opportunities we discussed in chapter 6.

This scenario envisions entrepreneurs and venture capital firms that have been successful in software investments recognizing the opportunities and importance of investing and commercializing deep science innovations and stepping up investment in such ventures. We can see early signs of this already happening in Silicon Valley. It remains to be seen, however, if the investments being made today will help turn back the tide toward deep science venture deals.

In the turnaround scenario, federal policymakers continue to forge ahead with new laws and regulations that foster greater access to capital for early-stage companies developing innovative products and applications with roots in deep science (as discussed in chapter 5). Constructive developments in U.S. capital markets, in turn, help reinvigorate and revitalize the innovation ecosystem and the important link among government, academia, venture capital, and deep science that fueled economic prosperity in the latter half of the twentieth century. Venture capital and well-functioning capital markets are critical ingredients to the healthy functioning of the deep science innovation ecosystem (highlighted in chapter 1).

As a sign that this scenario may already be underway in the venture investing world, Eric Ries, author of entrepreneurship manifesto *The Lean Startup*, is boldly venturing to launch what he calls a "long-term stock exchange" (LTSE). The LTSE would address the myopic focus on quarterly earnings and seek to encourage investors and companies to make better decisions for the years ahead. He notes that the most common conventional wisdom prevailing today is that going public means the end of a company's ability to innovate. He notes that many Silicon Valley startups are no longer thinking of IPOs and are being told not to go public.

Ries has assembled a team of about twenty engineers, finance executives, and attorneys and has raised a seed round from more than thirty investors to back the LTSE. Among the backers are venture capitalist Marc Andreessen, technology evangelist Tim O'Reilly, and Aneesh Chopra, the former chief technology officer of the United States. Discussions with the U.S. Securities and Exchange Commission have begun, but launching the LTSE could take several years. Whether the LTSE will come to fruition remains to be seen. The mere thought of a need for such a stock exchange is symptomatic of the issues described in our book.

Innovation is a core competency of a dynamic, thriving economy that produces millions of jobs and billions of dollars in income and wealth over time. Economies driven by innovation lead— whereas those that are not, follow. In this scenario, the return of venture investments to deep science becomes a primary catalyst for a resurgence of American economic dynamism, with positive effects on employment, income, wealth, and living standards.

It remains to be seen which scenario will play out in the future, or whether we will end up with something in between the two. Being rational optimists, we are inclined to believe that deep science venture deals will make a strong comeback in coming years. But there is no guarantee. A great deal of recalibration is needed in Silicon Valley, Washington, DC, and elsewhere in the United States to bring the turnaround scenario to fruition. We were impelled to spend the time and effort writing this book in the hope that we could help stimulate a constructive dialogue among investors, entrepreneurs, business leaders, scientists, educators, and policymakers regarding how best to address the issues discussed herein.

Appendix 1

The Case of D-Wave Systems

MODERN DIGITAL COMPUTERS USED by consumers, businesses, and government agencies are rooted in the inventions of quantum mechanics: the transistor, the integrated circuit, and the microprocessor. Although these devices employ some quantum effects, such as incoherent tunneling, they are not considered true *quantum* machines. To qualify as such, they must harness one or more quantum *coherent* phenomena, including superposition, interference, entanglement, and co-tunneling, which are among the most mysterious and counterintuitive properties associated with the quantum realm. Quantum science has created a vision of computer architecture that is fundamentally different from conventional digital computing technology: a *quantum computer*.

D-Wave Systems, Incorporated, manufactures a specialized type of quantum computer, called a *quantum annealer*, and has already sold several generations of them to cutting-edge technology customers, including Lockheed Martin; Google, NASA, and the Universities Space Research Association (USRA), for their Quantum Artificial Intelligence Lab; and the Los Alamos National Laboratory.

The story of how D-Wave came to be, how it chose to make quantum annealers instead of general-purpose quantum computers, and how deep science–based venture investing kept it afloat is one of the best examples of how venture capitalism can shatter preconceptions and overcome academic prejudices that hold technology back.

The Birth of Quantum Computing

The concept of a quantum computer has been around for decades, extending back to the 1980s and the creative speculations of Nobel laureate and Caltech professor Richard Feynman[1] and University of Oxford professor David Deutsch.[2] In the early 1980s, Feynman observed that it appeared impossible to simulate the evolution of a general quantum system on a classical computer efficiently and so proposed a basic model for a quantum computer that would be capable of such simulations. The field languished for the better part of a decade because it was unclear at the time how a quantum computer could detect and correct the errors that would arise during its operation, since classical error-correction methods all involved reading bits to detect errors. This would not work for quantum computers, because *reading* the qubits would necessarily change their state. Although efficient quantum simulation was clearly of interest to physicists, it was not immediately recognized as a sufficiently high-value application to warrant the massive investment necessary to make quantum computing a reality.

The situation changed dramatically in the mid-1990s when Bell Labs physicist Peter Shor discovered a key algorithm, today known as Shor's algorithm, which showed how a quantum computer could factor large composite integers and compute discrete logarithms efficiently.[3] Fast-integer factorization is the key to breaking the RSA public key cryptosystem, and fast computation of discrete logarithms is the key to breaking the elliptic curve public key cryptosystem. Shor's algorithm sparked the interest of the intelligence community and brought in significant funding for quantum computing. Shor also overcame the "error uncorrectability" objection to quantum computing by devising a scheme that circumvented the need to read the qubits directly in order to correct them.[4] Other key algorithms were discovered shortly thereafter, including Grover's algorithm for performing unstructured quantum search,[5] Cerf, Grover, and Williams's algorithm for structured quantum search,[6] and Abrams and Lloyd's algorithm for estimating eigenvalues of complex molecules.[7] The field of quantum computing was poised to explode.

Quantum Effects and Their Role in D-Wave's Quantum Computer

Quantum mechanical computers are distinguished from classical digital computers by the physical effects available to them at their lowest level of operation. In particular, quantum computers can harness *quantum* physical effects, such as superposition, interference, entanglement, and co-tunneling, which are simply not available to conventional computers no matter how sophisticated they may be. These physical effects allow quantum computers to solve problems in fundamentally new ways. Thus, quantum computers are not merely *quantitatively faster* than classical computers, but also *qualitatively different* from them. This sort of transformation was unprecedented in the history of computer science. Until the invention of quantum computers, all improvements in computer technology made machines that were formal equivalents to their predecessors. So, what is the nature of these new quantum effects, and what role do they play in computation?

Whereas classical digital processors manipulate *bits* (which can be 0 *or* 1), a quantum processor manipulates qubits, which can be in a *superposition*, or blend, of partly 0 *and* partly 1 *simultaneously*.

Superposition allows a quantum computer with a single one-qubit memory to hold two-bit string configurations simultaneously. Similarly, a quantum computer with a two-qubit memory can hold four-bit string configurations simultaneously (namely, 00, 01, 10, and 11). Likewise, a quantum computer with a three-qubit memory can hold eight-bit string configurations simultaneously (namely, 000, 001, 010, 011, 100, 101, 110, and 111).[8] One can see that a quantum computer with an n-qubit memory can hold a superposition of 2^n; that is, exponentially large numbers. This is noteworthy because when n is as small as 300, the number of components in the superposition (i.e., 2^{300}) is greater than the total number of fundamental particles in our entire universe! No past, present, or future classical memory could ever represent so many bit string configurations explicitly simultaneously, even if it was built from all available matter in our universe.

Stranger still, if it were possible to perform an operation on n bits in a certain amount of time, the quantum computer could perform that same operation on all 2^n components of the superposition in the same amount of time. The art of the quantum algorithm designer is to shape the quantum mechanical evolution of an initially equally weighted superposition so that the relative contribution of those bit string configurations corresponding to solutions to a desired problem are enhanced, whereas those corresponding to nonsolutions are diminished. This enhancement and diminishment in relative contribution occurs because of quantum mechanical interference between the many different computation pathways pursued, in quantum parallel, by the quantum computer. Trying to match quantum computers with classical parallelism can only fail.

Superposition is not a cure-all, however. Nature exacts a high price for the power to create and manipulate massive superposition states. Any attempt to read the superposition state at the end of the quantum computation will return only one of the many bit string configurations held within the state. And which bit string configuration is returned is impossible to predict in advance and can only be estimated probabilistically. This means that quantum computers are *necessarily* nondeterministic computers, and that the same quantum computation repeated several times will typically not return the same answer each time. We certainly see this behavior within the D-Wave quantum computer, which is usually used in a mode of resubmitting and resolving the same problem multiple times to obtain a set of bit string configurations each having some associated solution quality. When the D-Wave system is used as an optimizer, the bit string configuration of best quality is the only one returned. However, when the D-Wave system is used as a sampler, all bit strings found, together with their associated qualities, are returned. Typically, these correspond to a diverse sample of different solutions, or near-solutions, to the problem posed.

Other quantum phenomena are even more counterintuitive and have confounded even the greatest scientific minds, including Albert Einstein. For example, in a conventional computer, if we performed an operation on one subset of bits, we would not expect that act to change the values of the other untouched bits. This is not so with a quantum computer! As a quantum computer runs a quantum algorithm, quantum interference typically gives rise to superposition states of the quantum memory register that are impossible to factor into a definite state for each qubit within it.

Such superposition states, which are said to be "entangled," support the theory that reading one subset of qubits can profoundly change the values of the complementary, unread, subset of qubits. Moreover, the strength of the induced correlations between qubit values can be greater than what is possible classically. It is as if you and a friend agree to make some decisions, then go your separate ways and make your decisions independently from one another, but then later discover when you meet up again that you both made exactly the same decision even though you did not communicate your decisions to one another at the time you made them nor colluded on a preplanned set of decisions in advance! Such is the bizarre world of quantum entanglement. Through such entanglements, quantum computations can enforce correlations on qubit values that seem almost magical by classical standards.

The last quantum effect we will consider is quantum coherent tunneling. In classical physics, when confronted with a barrier of some kind, one typically expects to have to go around it. In the quantum realm, however, the opportunity arises instead to *tunnel through* such potential barriers. Moreover, unlike the incoherent tunneling that occurs inside conventional transistors, in which each electron tunnels independently of the others, in coherent tunneling, the set of qubits tunnels collectively from one energy minimum to a lower one. A good way to visualize how the D-Wave quantum computer works is to think of the problem to be solved as defining an implicit energy landscape in which better solutions correspond to lower points, with the optimal solution corresponding to the lowest point.

When the D-Wave quantum computer starts off, this landscape is made to be artificially flat so that all bit string configurations are initially equally likely. As annealing proceeds, the energy landscape of the problem is slowly imposed upon the initial energy landscape, and the qubits tunnel collectively from higher local energy minima into lower local energy minima. This process continues until, at the end of the quantum anneal, the quantum state representing the various surviving bit string configurations becomes highly localized to only the lowest-lying valleys. Reading the state of the quantum computer at this time reveals a bit string configuration that minimizes the energy function derived from the problem. By allowing the tunneling to occur coherently, the tunneling rate through the hills in the energy landscape can be made higher than the thermal hopping rate over the hills in the energy landscape (as occurs in a conventional computer), allowing the quantum machine to find the best solutions faster than would be possible if it had access only to classical physical effects or incoherent quantum effects.

The motivation for making a computer that harnesses quantum effects lies in the computational complexity advantage it confers. By harnessing quantum effects, a quantum computer can run certain algorithms *natively* that can only be *simulated inefficiently* classically. This parallels Feynman's original observation that classical computers cannot simulate general quantum systems efficiently. For factoring composite integers and estimating ground-state eigenvalues of complex molecules, an exponential speedup appears attainable. For a vastly larger, and arguably more commercially relevant, class of problems, including quantum search and solving NP-hard problems, a polynomial speedup appears attainable. Thus, with a quantum computer comes the potential for solving problems that do not yield to conventional computing approaches (figure ap1.1).

Focus areas for D-Wave quantum computing today include machine learning, artificial intelligence, computer vision, robotics, natural language understanding, computational finance, cryptology,

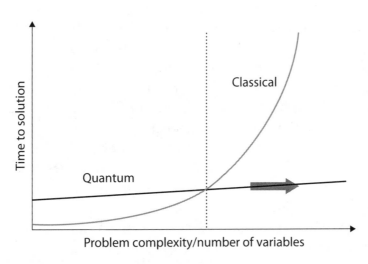

FIGURE AP1.1 The benefits of quantum computers versus classical computers. *Source*: D-Wave Systems.

software verification and validation, sentiment analysis, and quantum simulation. The deep science–related venture investment opportunity is therefore the potential to solve high-value computational problems across major market segments faster and better than is possible using conventional computers.

The Founding of D-Wave Systems

D-Wave's origins as a company were somewhat unusual. In 1999, Geordie Rose was given a homework assignment in a class he was taking at the University of British Columbia taught by legendary Canadian investor Haig Farris. Farris set his students the task of writing a business plan for a hypothetical company in any field—as long as they thought out of the box. Rose had just read the book *Explorations in Quantum Computing*, by Colin P. Williams, and, with quantum computing fresh in his mind, he chose to base his business plan on a hypothetical quantum computing company. The idea was to raise a pool of money from investors and use it to fund strategic R&D projects around the world in exchange for the intellectual property (IP) rights to the technology, with the hope of one day selling the IP portfolio to an anticipated technology company wanting to develop quantum computers commercially. Farris liked Rose's business plan so much that he offered to provide seed funding for the company. One of the technologies Rose thought promising was quantum computing based on d-wave superconductors, and so the company was called D-Wave Systems. The name stuck, even though the basis for its technology has now moved on to lower-temperature superconductors.

Geordie Rose, Haig Farris, Bob Wiens, and Alexandre Zagoskin co-founded D-Wave Systems, Incorporated, in Vancouver, British Columbia, in 1999, based on Rose's intellectual property aggregation strategy. The company began executing its business model, cemented strategic collaborations with various universities, and began accumulating IP. By 2004, progress was sufficient to entice the *Harvard Business Review* to write a case study on D-Wave's strategy.[9] In this phase, D-Wave learned the relative strengths and weaknesses of every approach to realizing quantum computing hardware proposed up to that date. This positioned the company perfectly to consider taking the leap to quantum hardware development.

Indeed, even in the *Harvard Business Review* case study, Rose openly speculated on whether the company should continue as an IP aggregator or switch to building its own quantum computer. The decision was made by the end of 2004, when Eric Ladizinsky joined D-Wave to help Rose make the transition from an IP play into a technology manufacturer. Ladizinsky's role was so pivotal, and the strategic change so great, that D-Wave honored Ladizinsky with the title of cofounder in recognition of his contribution to the new D-Wave. Since then, D-Wave has grown leaps and bounds and strives to exploit the deepest insights of physics and computer science to design quantum computers capable of taking on some of the world's hardest and most important challenges. Its customers include Fortune 500 companies, government agencies, research laboratories, and universities, who all welcome the opportunity to access quantum computing technology and to probe, program, and patent potential use cases at the earliest opportunity.

Financing Rounds

The path to commercializing D-Wave's quantum computer has been relatively long compared to the majority of venture investments in typical technology plays, like social media software startups. Its headquarters have moved out of downtown Vancouver into the suburb of Burnaby, British Columbia, and it now has a U.S. office in Palo Alto, California, in the heart of Silicon Valley.

Thus far, D-Wave has raised C$173.3 million through seven rounds of institutional financing in the private market. The financing rounds are as follows: Series A (based on Class A Series 1 stock): December 21, 2001 to December 17, 2002; Series B (based on Class A Series 2–4 stock): May 16, 2003 to July 30, 2004; Series C (based on Class B stock): April 19, 2006 to June 16, 2006; Series D (based on Class C/D stock): January 23, 2008 to September 30, 2008; Series E (based on Class E/F Series 1 stock): November 24, 2010 to February 28, 2011; Series F (based on Class E/F Series 2 stock): June 1, 2012 to March 22, 2013; and Series G (based on Class G/H stock): June 27, 2014 to December 23, 2014. Within a round, there are multiple closings; the aforementioned dates show the first and last closings of each round.

Series A Financing Round

D-Wave's first institutional Series A financing round occurred between December 2001 and December 2002, closing with a total of C$3 million raised. Investors included GrowthWorks and BDC Capital.

Post–Series A Funding Evolution

The funding allowed the company to advance its IP accumulation strategy and begin exploring novel designs for its own superconducting qubits. At this time, D-Wave was focused on gate model quantum computing, as the adiabatic model was new, not well understood, and had even been greeted with somewhat derisive comments by some in the mainstream quantum computing community.

Series B Financing Round

D-Wave raised its Series B round between May 2003 and July 2004, closing the round with a total of C$11 million raised. Participating in the round were BDC Capital, Draper Fisher Jurvetson, and the British Columbia Investment Management Corporation (BCIMC).

Post–Series B Funding Evolution

The Series B funding provided D-Wave with sufficient resources to take on the challenge of building its own superconducting processors to a design inspired by the adiabatic model of quantum computing. The company began searching for a suitable candidate to spearhead a new fabrication-oriented effort. Rose asked Colin P. Williams, who had broad visibility across the quantum computing community, for suggestions. Williams recommended hiring Ladizinsky, who in 2000 had won a $10 million project with the Defense Advanced Research Projects Agency (DARPA) to fund a mini–Manhattan Project to build a superconducting gate-model quantum computer. Ladizinsky and Rose clicked instantly, and Ladizinsky came into D-Wave with gusto. He rapidly bought into the vision of building an adiabatic-model quantum computer, assembled a world-class team of superconducting quantum electronics engineers to design it, and cemented a deal with the NASA Jet Propulsion Laboratory at the California Institute of Technology to begin fabricating prototypes. Ladizinsky also recognized the need to cultivate a Silicon Valley startup culture within D-Wave. He actively promoted greater internal communications, processes for rapidly attaining internal alignment, and instilled healthy strategies for conflict resolution. These changes enabled D-Wave to become more nimble, more collaborative, and more tightly focused on the single-minded pursuit of building an adiabatic quantum processor. Moreover, the new culture valued rapid prototyping, open communication, internal alignment, and teamwork.

In January 2005, D-Wave promoted Jeremy Hilton from his role as director of intellectual property to vice-president of processor development. In his new capacity, Hilton assumed responsibility for the development of D-Wave's quantum processors. This included providing technical leadership in the management of the core team of physicists and quantum engineers developing the D-Wave hardware and the design, fabrication, modeling, and testing of superconducting devices and processors. Under Hilton's leadership, the D-Wave product would eventually mature from a sixteen-qubit lab prototype through three generations of commercially released quantum processors, culminating in the one-thousand-plus-qubit D-Wave 2X in 2015.

In May 2005, D-Wave hired Bill Macready as vice-president of software engineering. Macready began recruiting a team of highly talented computer scientists to complement D-Wave's strong team of in-house physicists. The software team created application program interfaces (APIs) for interacting with the D-Wave system from Matlab, C/C++, and Python. They also developed tools for converting a variety of problems into quadratic unconstrained binary optimization (QUBO) problems, translating arbitrarily structured QUBOs into ones that conformed to the interconnected topology of the D-Wave machine and mapping such properly structured QUBOs to Ising problems, which could then run natively on the D-Wave processor, ultimately making it easier to program the D-Wave machine. The team also began developing applications that showed how to use the hardware to solve discrete combinatorial optimization, constraint satisfaction, and discrete sampling problems. Specific examples included factoring composite integers, computing Ramsey numbers, performing constraint satisfaction as inverse verification, solving assignment problems, and optimizing molecular matching.

Another key hire was made in December 2005, when D-Wave hired Tanya Rothe as general counsel and director of intellectual property. Rothe set to work transitioning D-Wave's IP strategy from that of aggregation to a more strategic approach based on capturing and protecting internally generated IP broadly and in a cost-effective manner. This strategy was immensely successful and led to IEEE Spectrum ranking D-Wave's patent portfolio the fourth most important in the computer systems category in December 2012.[10]

Series C Financing Round

In April 2006, D-Wave completed a Series C round of venture financing that raised C\$14.7 million. Participating in the round were Working Opportunity Fund (GrowthWorks), BDC Capital, Draper Fisher Jurvetson, Harris & Harris Group, and BCIMC.

Post–Series C Funding Evolution

In September 2006, D-Wave hired Dave Pires as vice-president of hardware engineering, giving him the task of turning D-Wave's technology into a deployable product. This included developing refrigeration systems capable of cooling the processor and associated electronics to millikelvin temperatures, designing analog control lines capable of carrying signals from room temperature to the super-cool processor, and engineering an ultra-low-noise and low-magnetic-field environment for the quantum processor. With such improvements, made possible by the Series C funding, by 2007, D-Wave had designed its first sixteen-qubit processor based on superconducting niobium qubits. On February 13, 2007, the "Orion" quantum processor was revealed to the world in a live demonstration at the Computer History Museum in Mountain View, California. Before a large audience, Rose demonstrated the processor solving a molecular matching problem, a seating assignment problem, and a Sudoku problem. The demonstration also illustrated the feasibility of interacting with a D-Wave quantum computer via an Internet connection and standard computer interface. The event was met with strong skepticism by some in the mainstream quantum computing community who had neither heard of D-Wave nor thought much about the adiabatic model of quantum computing beforehand. Nevertheless, and despite the negative publicity that ensued, the Computer History Museum event turned out to be a seminal moment for D-Wave, as it caught the attention of Ned Allen, chief scientist at Lockheed Martin, and Hartmut Neven, of Google, who would both later play pivotal roles in facilitating purchases of D-Wave's quantum computers.

Despite the success of Orion as a functional quantum processor, its design was not scalable, owing to the need for excessive numbers of control lines. Subsequently, D-Wave redesigned its architecture to be scalable to any size by incorporating integrated programmable magnetic memory into the processor, which dramatically reduced the number of lines needed to be fed into the dilution refrigerator housing the quantum processor.

Insider and Series D Financing Round

In January 2008, D-Wave raised C$17 million from insiders. In September, D-Wave announced the completion of a Series D round that brought in C$11.4 million from institutional investors including Goldman Sachs, International Investment and Underwriting, and Harris & Harris Group.

Post–Series D Funding Evolution

Subsequent to the Series D round, D-Wave began an informal collaboration with Google on quantum machine learning. This culminated in academic papers coauthored with Google that described how to train binary classifiers using quantum annealing.[11,12] In 2009, Google and D-Wave gave a joint presentation on their quantum machine-learning algorithm at the Annual Conference on Neural Information Processing Systems.[13] The processor had 128 qubits and used thirty-three thousand superconducting Josephson junctions etched onto a four-by-seven-millimeter die. This basic design has gone through several cycles of refinement. The new flagship D-Wave 2X product now boasts over one thousand qubits and 128,000 Josephson junctions and is believed to be the most sophisticated superconducting processor of any kind ever made.

In September 2008, Warren Wall joined D-Wave as chief operating officer (COO). Wall was formerly COO of Electronic Arts Canada and acted as interim D-Wave CEO from November 2008 until September 2009.

In September 2009, D-Wave successfully recruited Vern Brownell as CEO to lead the company through its transition from research into the world's first commercial quantum computing company. Prior to joining D-Wave, Brownell had been CEO of Egenera, a pioneer of infrastructure virtualization, and chief technology officer (CTO) at Goldman Sachs, where he and his staff of 1,300 were responsible for worldwide technology infrastructure.

In spring 2011, under Brownell's tenure, D-Wave sold its first quantum computer to Lockheed Martin. The sale included a multi-year contract, the installation of a 128-qubit D-Wave One system, a maintenance agreement, and associated professional services to enable the two companies to collaborate on applying quantum annealing to some of Lockheed Martin's most challenging computational problems. In the fall of 2011, D-Wave deployed the first operational quantum computer system at an academic institution by installing a D-Wave One at the University of Southern California's Information Sciences Institute in Marina del Rey, California.

The first successful commercial sale of a D-Wave quantum computer system enabled the company to begin to attract talent across all levels of the organization. In summer 2012 Williams, author of the book that had inspired Rose to start D-Wave, joined the company to serve in a variety of capacities, including as director of business development and strategic partnerships. A year later, Bo Ewald, former president of Cray and CEO of Silicon Graphics, joined D-Wave to serve as its chief revenue officer. In 2014, Bill Blake, former CTO of Cray, joined D-Wave to serve as executive vice-president of R&D.

Series E Financing Round

In February 2011, D-Wave completed a Series E financing round that raised C$17.5 million. Investments were provided by the large existing investors as well as newcomer Kensington Capital Partners.

Post–Series E Funding Evolution

The Series E funding allowed the company to improve and scale up the 128-qubit D-Wave One system into the 512-qubit D-Wave Two. In addition to increasing the number of qubits, the qubits were also made smaller, while their energy scale was expanded, making it less likely that once the system found the lowest energy state it would later escape from it. Due to the shrinking qubits, the footprint of the 512-qubit "Vesuvius" chip was about the same as that of the 128-qubit Rainier chip, but its performance was enhanced dramatically. In addition, the Series E funding allowed D-Wave to improve the method by which problems could be posed to the chip. In the old Rainier design, the act of programming the chip introduced a small amount of heat, which raised the temperature of the chip. Before quantum computation could commence, users had to wait for the chip to cool to the base temperature of the dilution refrigerator, which added to the computation time. With the bolus of Series E funding, D-Wave was able to redesign the input/output (I/O) system in a manner that reduced the heat delivered to the chip when programming it. This allowed the wait time between programming the chip and running the problem to be cut dramatically, thereby enhancing performance. Indeed, one of D-Wave's primary strengths has been to consistently and repeatedly identify, prioritize, and overcome engineering challenges that have obscured the underlying performance of quantum computing.

In May 2011, a major milestone was reached when D-Wave published a paper in *Nature* entitled "Quantum Annealing with Manufactured Spins."[14] This paper demonstrated compelling experimental evidence that the dynamics at play inside the D-Wave processor matched that of quantum annealing, rather than classical thermal annealing. Moreover, it showed that the device could be configured, in situ, to realize a wide variety of different spin networks. The authors demonstrated that D-Wave had built a true programmable quantum annealer, which added considerably to the company's credibility.

In 2012, buoyed by new designs for the chip and its accompanying I/O system, D-Wave expanded its team and transitioned manufacturing to a commercial-grade complementary metal–oxide–semiconductor (CMOS) foundry, leading to lower device variance, a higher qubit yield, and improved chip quality. This transition was made possible by D-Wave showing that it could introduce niobium into a mainline semiconductor process without it contaminating the equipment, which was no small feat, given that niobium is an element not typically used in CMOS manufacturing.

Series F Financing Round

In March 2013, D-Wave closed a C\$34 million Series F equity financing round to further advance the development of its quantum computer business. The round included noted investors Bezos Expeditions, the personal investment fund of Amazon founder Jeff Bezos, and In-Q-Tel, a strategic investment firm that delivers innovative technology solutions in support of the missions of the U.S. intelligence community. The Series F round also involved participation from existing institutional investors, including BDC Capital, Draper Fisher Jurvetson, Goldman Sachs, GrowthWorks, Harris & Harris Group, International Investment and Underwriting, and Kensington Capital Partners.

Post–Series F Funding Evolution

In spring 2013, D-Wave announced that its 512-qubit D-Wave Two quantum computer had been selected for installation at the then newly established Quantum Artificial Intelligence Laboratory (QuAIL) at the NASA Ames Research Center. QuAIL was created as a collaborative initiative among Google, NASA, and USRA, with the mission of exploring the potential for quantum computing to advance the state of the art in artificial intelligence and machine learning. The deal built upon a successful research collaboration with Google on quantum machine learning that had begun in 2008 and on a study D-Wave had conducted with NASA in 2012 entitled "A Near-Term Quantum Computing Approach for Hard Computational Problems in Space Exploration."[15] In 2014, the D-Wave system was successfully installed in NASA's Advanced Supercomputing facility, and its administration was assigned to USRA. In the summer of 2014, USRA announced a request for proposals from people worldwide wishing to run problems on the D-Wave system. The first papers arising from such projects began appearing in October 2015.[16]

Another milestone was achieved in the spring of 2013 with the announcement of the publication of a peer-reviewed paper on D-Wave's quantum computer in *Nature Communications*.

The paper, entitled "Thermally Assisted Quantum Annealing of a 16-Qubit Problem," presented the results of the first experimental exploration of the effect of thermal noise on quantum annealing.[17] Using a feasible-to-analyze subsystem of sixteen qubits within a D-Wave processor, the experiments demonstrated that, for the problem studied, even with annealing times eight orders of magnitude longer than the predicted single-qubit decoherence time (the typical time it takes for environmental factors to start to corrupt the state of a qubit), the probabilities of performing a successful computation are similar to those expected for a fully coherent system. The experiments also demonstrated that by repeatedly annealing an open (i.e., coupled to its environment) quantum system quickly several times, rather than annealing a hypothetical closed (i.e., not coupled to its environment) system slowly once, quantum annealing can take advantage of a thermal environment to achieve a speedup factor of up to one thousand over the closed system. Similar results were later confirmed by a research group at the University of Southern California.[18]

In the summer of 2013, D-Wave announced it had passed a milestone in its IP strategy by having its one hundredth patent granted by the U.S. Patent and Trademark Office. And in December 2012, IEEE Spectrum rated D-Wave's patent portfolio fourth in its computer systems category,[19] just behind computing giants IBM, HP, and Fujitsu and beating out Cray and Silicon Graphics; this signaled a major accomplishment for the emerging company pioneering transformative technology based on deep science.

By now, media interest in D-Wave was picking up. In February 2014, D-Wave's quantum computer was featured in a *Time* magazine cover story, "The Quantum Quest for a Revolutionary Computer."[20]

Also in winter 2014, D-Wave was named to *MIT Technology Review*'s 2014 list of the "Fifty Smartest Companies" and became recognized as a quantum technology leader among the world's most innovative companies.[21] Companies included in the *MIT*

Technology Review list were said to have demonstrated original and valuable technology in the past year and were bringing that technology to market at significant scale while strongly influencing their competitors, thereby representing the most disruptive innovations most likely to change people's lives.

In April 2014, a team of scientists from the University of Southern California and University College London posted a preprint, which was published in a peer-reviewed journal in 2015, showing that the D-Wave processor fit an open-system quantum dynamical description extremely accurately, even in the presence of thermal excitations and despite the chip having a small ratio of single-qubit coherence time to annealing time.[22] This paper effectively silenced critics from IBM and Berkeley who had proposed various classical models to explain what the D-Wave device was doing.[23,24] This paper showed, explicitly, that *none* of those models fit the D-Wave device's behavior.

One of the seminal moments in D-Wave's history came in May 2014, when D-Wave published a paper entitled "Entanglement in a Quantum Annealing Processor," which provided conclusive and irrefutable experimental evidence for the presence of quantum entanglement within the company's computers.[25] Not only did the paper show the presence of entanglement throughout the critical stages of quantum annealing, it also showed that such entanglement was persistent at thermal equilibrium and was occurring at a world-record level for superconducting qubits. Thus, the paper showed that entanglement is not as fragile and fleeting as many D-Wave critics had believed, and the fact that the research was published in *Physical Review X* was a significant milestone for D-Wave and a major step forward for the science of quantum computing.

Later, in November 2014, a team of scientists from Google and D-Wave published a paper showing that not only were quantum effects present in the D-Wave Two processor, but that they also played functional role in the computation it performed.[26]

Similarly, other studies showed that, contrary to a belief held by many gate-model quantum computing researchers, it *is* possible to error correct a D-Wave quantum annealer.[27, 28] This latter realization took away much of the steam from critics who claimed the D-Wave architecture would not scale.

More good news came in the summer of 2014, when D-Wave announced partnerships with two startup companies aiming to write software applications that would build upon D-Wave technology. One of them, DNA-SEQ, uses D-Wave's machine learning capabilities, in conjunction with its own proprietary biotechnology, to find drugs that would best fight the cancer of individual patients on a case-by-case basis. Another company, 1QBit, is writing software for financial applications that is ideally suited to D-Wave's quantum annealing architecture.

Following the Series F C$34 million capital raise, D-Wave was successful in recruiting Bill Blake, former CTO of Cray, as executive vice-president of research and development, responsible for hardware engineering, processor development, software, and fabrication process design functions. Tragically, Blake died unexpectedly shortly after joining the company. However, in his short tenure, he was instrumental in shaping the company's strategy in machine learning and developing the system architecture needed to integrate D-Wave into a traditional high-performance computing environment.

Series G Financing Round

In December 2014, D-Wave announced the closing of an additional C$61.7 million of equity funding in a Series G financing round to advance and scale the company's quantum computing technology and accelerate the development of quantum computing software. Investors in the round included Goldman Sachs, BDC Capital, Harris & Harris Group, and Draper Fisher Jurvetson.

Post–Series G Funding Evolution

In the summer of 2015, D-Wave announced the D-Wave 2X, a one-thousand-plus-qubit quantum processor. This new processor, comprised of over 128,000 Josephson junctions, marked a major milestone for the company and, at present, is by far the most complex superconducting integrated circuit ever successfully yielded. The one-thousand-plus-qubit processor represents a major technological and scientific achievement that allows significantly more complex computational problems to be solved than was possible on earlier quantum computers. In addition to scaling the processor to beyond one thousand qubits, the new system incorporates other major technological and scientific advances and operates at a temperature below 15 millikelvin, which is 180 times colder than interstellar space and very close to absolute zero.

In the fall of 2015, D-Wave announced that it had entered into a new agreement with Google, NASA, and USRA to allow for the installation of a succession of D-Wave systems at the NASA Ames Research Center in Moffett Field, California. This agreement supports and extends the QuAIL collaboration among Google, NASA, and USRA. The partnership is dedicated to the continued study of how quantum computing can advance artificial intelligence, machine learning, and the solution of difficult optimization problems. The new agreement enables Google and its partners to keep their D-Wave system at the state of the art for up to seven years, with new generations of D-Wave systems to be installed at the Ames Research Center as they become available. Other key announcements included the sale of a one-thousand-plus-qubit D-Wave 2X system to Los Alamos National Laboratory and a multiyear agreement with Lockheed Martin, which includes an upgrade the company's 512-qubit D-Wave Two quantum computer to the D-Wave 2X system.

Also at this time, 1QBit published a paper with an analyst at Guggenheim Partners showing how to apply the D-Wave machine to solve an optimal trading trajectory problem.[29] The paper studied a multiperiod portfolio optimization problem using the D-Wave Systems quantum annealer, derived a formulation of the problem, discussed several possible integer encoding schemes, and presented numerical examples that showed high success rates.

Summary: Into the Quantum Computing Age

The famous Nobel Prize–winning physicist Richard Feynman started the quest for quantum computing about thirty years ago. Since then, there have been multiple attempts to build a quantum-computing device, both in academia and in industry. Until recently, the academic community did not expect a quantum computer to be built within the next fifty years. That forecast has changed, thanks in large part to D-Wave and the venture investors who have backed the company.

The number of technical breakthroughs delivered by D-Wave since the company's founding has been staggering. Total invested capital in the company to date is around C\$176 million from a blue-chip investor base. Over the past decade, D-Wave's deep scientists have authored more than seventy peer-reviewed papers published in prestigious scientific journals, including the *Journal of Computational Physics*, *Nature*, *Physical Review*, *Quantum Information Processing*, and *Science*.

Since the company's founding in 1999, D-Wave has been assiduously building a quantum computing knowledge and intellectual property base. The company's patent portfolio today stands at over 165 issued patents worldwide, covering all aspects of its technology.

Additionally, D-Wave is pioneering more than just quantum computing; the company is also accumulating experience with new paradigms—such as superconductivity—that could keep Moore's law going. D-Wave's superconducting processors have advantages even outside of being able to perform quantum computing, such as the fact that they release no heat at all. There is also the potential for the technology to improve further, perhaps being able to carry out the next paradigm that will continue Moore's law when traditional transistors reach their physical limits.

Led by CEO Vern Brownell, today D-Wave employs over 120 people and plans to expand its hiring in the future. The company is strategically moving toward integrating quantum computing into the mainstream computing infrastructure and using its transformative computing technology to solve a range of challenging real-world problems across a variety of segments and applications. Over the past five years, several of the most prestigious organizations in the world, including Google, Lockheed Martin, NASA, USRA, and the University of Southern California, have used D-Wave's quantum computing systems and conducted pioneering research in machine learning, discrete optimization, and physics. Key industry applications of D-Wave's quantum computers include everything from solving complex optimization problems found throughout industry to accelerating machine learning to gain critical insights from enormous data sets. Partnerships with third-party software developers like 1QBit are fostering the development of a quantum software ecosystem that will help to promote system sales.

With the ascent of D-Wave's quantum computers, it seems clear that we have entered a new era of computing. D-Wave's quantum processor demonstrates both local entanglement and coherent tunneling, phenomena that are part and parcel of the quantum realm and not of classical physics. Meanwhile, other companies, such as BBN, IBM, Microsoft, and NEC, are doing pioneering work in the field of quantum computing, but this research is focused on alternative methods of quantum computation related to building a universal quantum computer in the gate-model architecture or the topological architecture. There are four or five different approaches to building a quantum computer today, but D-Wave's use of superconducting flux qubits in a quantum annealing architecture is the only one that currently works at scale.

Despite D-Wave's remarkable technological breakthroughs over the years, the jury is still out on the scalability of D-Wave quantum computers. However, the lessons learned from analyzing the 512-qubit D-Wave Two processor have led the quantum computing community to a deeper understanding of the sorts of problems with which one would expect to see a computational advantage for quantum annealing. In particular, the benchmarking that was done on the 512-qubit D-Wave Two focused on the scaling of the time taken to reach a global optimum. It is now understood that if problems are pathologically easy, they are unlikely to benefit from solution via quantum annealing, as even poor classical solvers can race to a solution very quickly.[30] Moreover, it is now also understood that the time to achieve a global optimum is especially sensitive to misspecification in the parameters required to pose a computational problem to the chip. When one respects the precision with which it is feasible to specify problems to the D-Wave processor, the performance of the D-Wave machine is much better.[31] This therefore gives a truer sense of the potential for quantum annealing.

In August 2015, the first performance benchmarking results for the one-thousand-plus-qubit D-Wave 2X appeared.[32] This study learned from the inadequacies of the time-to-solution (TTS) metric and focused instead on a new time-to-target (TTT) metric. The TTT metric asks not how long it takes the machine to find a global optimum, but rather how long it takes it to find a solution having a specific target quality. This metric is more aligned with real-world problems in which one typically does not know, or does not even know how to recognize, the global optimal solution in advance. With respect to the TTT metric, the D-Wave 2X was found to be up to six hundred times faster in raw annealing time than the best competing classical solvers on one problem class tested.

D-Wave's current quantum processor has more than one thousand qubits, allowing it to solve optimization problems with up to one thousand variables in single machine instruction. However, larger problems can be tackled by using the D-Wave system iteratively on up to one-thousand-variable subsets of a larger problem. Key areas of future growth for the company include deeper basic R&D, more processor development, greater systems integration, and wide-ranging software engineering.

D-Wave offers several APIs to allow customers to program its systems. APIs/compilers for Matlab, C/C++, and Python currently exist, and those for Julia and Mathematica are in the works. Qubit yield in the next-generation processor is expected to be around 98 percent, which is expected to make it easier to embed problems onto the chip, whose structure does not match its native form. The company is also working on more advanced algorithms and real-world applications, such as neural networks to harness quantum effects to accelerate deep learning and artificial intelligence.

While D-Wave's quantum processor is impressive, the company will need to demonstrate, through its partnerships and ongoing research, that its quantum computer has advantages over conventional digital computers. The potential exists for D-Wave to demonstrate the superiority of quantum computing in the marketplace, but there is a steep learning curve associated with quantum computing. In this regard, there is tremendous benefit to D-Wave of collaborating with organizations such as Google, Lockheed Martin, and NASA and universities such as Harvard, University College London, and the University of Southern California.

In contrast with the recent venture investing fascination with quick-turn social media companies, what D-Wave is trying to do is incredibly complex from scientific, technological, and manufacturing standpoints. What makes such a project especially challenging is that the probability of failure tends to grow as the square of complexity. Getting the right financial resources in place adds yet another level of complexity, turning D-Wave Systems into

Financing Rounds and Selected Corporate Milestones

1999	2002–2005	2006–2009	2010–2012	2013	2014–2015
Company founded	Series A Round: C$3 m Series B Round: C$14 m Eric Ladizinsky joins company Bill Macready hired as VP of software engineering Tanya Rothe hired as general counsel and director of intellectual property	Series C Round: C$14.7 m Insider Round: C$17 m Series D Round: C$11.4 m Dave Pires hired as VP of hardware engineering D-Wave begins informal collaboration with Google on quantum machine learning Warren Wall hired as COO and Vern Brownell as CEO	Series E Round: C$17.5 m First D-Wave 52-qubit quantum computing system sold to Lockheed Martin "Quantum Annealing with Manufactured Spins" published in *Nature* Colin P. Williams, noted quantum computing expert, author, scientist, and lecturer, joins company	Series F Round: C$34 m D-Wave Two quantum computer selected for new quantum artificial intelligence initiative; system to be Installed at NASA's Ames Research Center (with Google) "Thermally Assisted Quantum Annealing of a 16-qubit Problem" published in peer-reviewed *Nature Communications* Bo Ewald, former president of Cray and CEO of Silicon Graphics, hired as chief revenue officer	Series G Round: C$61.7 m D-Wave featured in *Time* magazine cover story on quantum computing 1,000+ qubit processor developed Joins the NSF Center for Hybrid Multicore Productivity Research (CHMPR) Research validates quantum entanglement in D-Wave's computers Announces multiyear quantum computing agreement with Google, NASA, and USRA Sale of 1,000+ qubit D-Wave 2X system to Los Alamos National Laboratory Announces multiyear quantum computing agreement with Lockheed Martin that includes upgrade from 512-qubit to 1,000+ qubit system

FIGURE AP I. 2   The D-Wave Systems timeline. *Source:* D-Wave Systems.

something of a Manhattan Project. It is a huge undertaking, and one that pushes the boundaries of science and technology. For now, D-Wave is the undisputable leader of scalable quantum computing and a poster child for pioneering the commercialization of transformation technology based on deep science.

D-Wave's evolution since the company's founding in 1999 is summarized in figure ap1.2.

Appendix 2

The Case of Nantero

JUST AS MOORE'S LAW says, for a long time scientists and engineers have found ways to double the number of transistors on integrated circuits every two years, significantly increasing their power and functionality over time. Many consumers, businesses, and venture investors take Moore's law for granted today. They expect their computers and electronic devices to become more powerful over time, delivering a bigger bang for the buck and enabling new business models (e.g., social media) and applications (e.g., Facebook, Uber).

As we migrate deeper into the quantum realm down to ten nanometers and beyond, scientists are questioning the ability of silicon to continue to power Moore's law. Current research suggests problems with silicon at the sub-seven-nanometer level—a level likely to be reached by the end of the decade. While predictions of an imminent end to the viability of Moore's law over the past decade have been off the mark, there is concern among scientists and engineers that we may be nearing the end of the silicon age.

Even Gordon Moore himself expressed concern about the viability of his law in the years ahead.[1]

While silicon may still be the dominant material for the production of microprocessors and semiconductors over the next five years, scientists and researchers are exploring other materials that are highly scalable down to one or two nanometers and would keep Moore's law chugging along for the foreseeable future. One of those that are promising is carbon nanotubes (CNTs).

CNTs are remarkably strong elastic cylinders of carbon atoms that bear a resemblance to a tube of rolled-up chicken wire. CNTs are tiny: One nanotube is just one-fifty-thousandth the diameter of a human hair. CNTs also have unique structural and electrical properties: They are 117 times stronger than steel and half the density of aluminum. They have electrical and thermal conductivity properties that make them useful for a host of applications in a wide range of industries, from aerospace and transportation to energy, water, and electronics. Since their discovery in 1991, CNTs have become a focus of active research in the United States and overseas. They are seen as a potential replacement for silicon in semiconductors in the future. The high tensile strength and thermal and electrical conductivity of CNTs make them very attractive for electronic device applications. These attractive properties enable performance breakthroughs through incorporation into existing and next-generation semiconductor products.

The Founding of Nantero

Greg Schmergel, Brent Segal, and Thomas Rueckes founded Nantero in 2001 in Woburn, Massachusetts, with Schmergel as CEO and Segal as COO. Rueckes is the inventor of the innovative nanoelectromechanical NRAM (a type of nonvolatile random-access memory) design and the company's CTO. Schmergel, Segal, and Rueckes assembled a scientific team of specialists in nanotechnology and semiconductors to develop Nantero's nanotechnology.

Nantero was launched to develop and commercialize Rueckes's NRAM invention—high-density, nonvolatile, random-access-memory-using CNTs. Nantero's founders envisioned NRAM as a universal memory device that would replace all current forms of memory, such as dynamic random-access memory (DRAM), static random-access memory (SRAM), and flash memory. The company's founders sought to develop a proprietary mode to manufacture NRAM and integrate it with standard semiconductor processes, which would help advance Nantero's CNT memory device into the market more quickly.

Nantero initially envisioned a market opportunity for its NRAM chip in excess of $70 billion per annum. The company's founders saw NRAM enabling instant-on computers and replacing DRAM and flash memory in devices like smartphones, tablets, laptops, MP3 players, digital cameras, and different types of enterprise systems. Other applications were envisioned in networking. Nantero's business model is akin to a fabrication-less ("fabless") semiconductor company. It is based on licensing the company's proprietary CNT technology and processes to manufacturers around the world, similar to ARM Holdings, which has had phenomenal success over the past decade.

Series A Financing Round

Nantero announced its first round of investment in October 2001 to begin developing its core NRAM and its technology and related processes. The Series A round raised $6 million. Investors in the Series A financing round included venture firms Draper Fisher Jurvetson, Harris & Harris Group, and Stata Venture Partners. Alex d'Arbeloff, chairman of MIT and founder of Teradyne, was another investor, and he was also added to Nantero's board of directors.

It is noteworthy that the financial environment in 2001 was not especially conducive to raising capital for emerging-technology companies—particularly companies pioneering the development of nanotechnology. In the year leading up to the completion of Nantero's Series A round, the Nasdaq stock index—a bellwether for emerging-technology stocks—had fallen precipitously, driven by a collapse of valuations in the previously high-flying dotcom segment.

Post–Series A Funding Evolution

In spring 2003, Nantero announced the achievement of a milestone. The company had successfully created an array of ten billion suspended nanotube junctions on a single silicon wafer. This development was significant, as it demonstrated that nanotubes could reliably be positioned in large arrays and scaled up to make even larger arrays. Nantero's process also resulted in substantial redundancy for the memory. This was a result of the fact that each memory bit depends not on one single nanotube, but upon many nanotubes that are woven akin to fabric. The highly conductive single-layer nanotube fabrics were seen to have an extensive range of applications beyond memory chips: transistors, sensors, and interconnects. Creating this vast array of suspended nanotubes using standard semiconductor processes brought Nantero closer to its goal of mass-producing NRAMs.

Nantero's innovative design for NRAM involved using suspended nanotube junctions as memory bits, with the "down" position representing bit 1 and the "up" position representing bit 0. Bits are switched between states through the application of electrical fields. Importantly, the wafer used to produce the NRAM was produced using only standard semiconductor processes, thus maximizing compatibility with existing semiconductor fabs. Nantero's proprietary method for achieving this result involved the deposition of a very thin layer of carbon nanotubes over the entire surface of the wafer. Lithography and etching were then employed to remove the nanotubes that were not in the correct position to serve as elements in the array.

During 2003, Nantero added Mohan Rao to its scientific advisory board. Rao was one of the world's leading very-large-scale-integration (VLSI) chip designers, having served previously as senior vice-president of the Semiconductor Group at Texas Instruments. He had extensive experience in semiconductors, holding over one hundred patents worldwide on aspects of memory, including SRAM, DRAM, and system-on-chip.

Series B Financing Round

In September 2003, Nantero announced the successful completion of a Series B round of financing. The company raised $10.5 million to further advance the development of its NRAM technology. The Series B round was led by Charles River Ventures, a venture firm with over three decades of experience in technology. Bruce Sachs and Bill Tai, both partners at Charles River Ventures, joined Nantero's board of directors. Returning institutional investors included Draper Fisher Jurvetson, Harris & Harris Group, and Stata Venture Partners.

Post–Series B Funding Evolution

Around the same time as its Series B capital raise, Nantero announced a collaboration with ASML, a leading global provider of lithography systems for the semiconductor industry. Nantero had developed a manufacturing process for NRAM that was completely complementary metal–oxide–semiconductor (CMOS) compatible. The collaborative work with ASML involved demonstrating that Nantero's proprietary process was achievable using conventional lithography equipment. The joint effort demonstrated that ASML's equipment could be employed with nanotubes using Nantero protocols and carrying out Nantero's proprietary manufacturing steps without any modifications. This was a major milestone for the company. It set Nantero on course to further advance the commercialization of its CNT memory devices.

In the spring of 2004, O. B. Bilous joined Nantero's board of directors. At the time, Bilous was chairman of the board of International SEMATECH, held fifteen patents, had published numerous articles, and had presented at many conferences and technical meetings in the area of semiconductor technology. Previously, Bilous had been vice-president of worldwide manufacturing for IBM in microelectronics.

During the summer of 2004, Nantero announced it had been granted a seminal patent covering CNT films and fabrics. The patent relates to a CNT film made of a conductive fabric of CNTs deposited on a surface. The CNT film can be used in semiconductors, wherein the film is deposited on a silicon substrate. The carbon nanotube film is a significant innovation enabling high-volume, cost-effective manufacturing of CNT-based devices and other products. With the addition of the new patent, Nantero's patent portfolio expanded to ten granted U.S. patents and a pipeline of more than forty additional patents pending.

Also that summer, Nantero announced a CNT development project with LSI Logic. The project involved the development of semiconductor process technology, advancing the effective use of CNTs in CMOS fabrication. At the same time, the company also announced a joint evaluation of CNT-based electronics with BAE Systems. This collaboration involved evaluating the potential to develop CNT-based electronic devices for use in advanced defense and aerospace systems. The project involved the R&D of a host of next-generation electronic devices that could be produced employing the unique properties of CNTs and leveraging

Nantero's proprietary methods and processes for the design and manufacture of CNT-based electronics.

In the fall of 2004, Nantero announced that it had received $4.5 million in funding to develop CNT-based radiation-hard, nonvolatile RAM with the Coalition for Networked Information and the Center for Applied Science and Engineering at Missouri State University. The radiation-resistant nanotechnology was aimed for use in space for U.S. defense purposes. In early 2005, Nantero announced it was actively seeking manufacturing partners in Asia and Europe to further advance the commercialization of NRAM in the consumer electronics segment. France, Germany, Italy, Japan, Korea, and the Netherlands were among the countries targeted for possible licensing of the NRAM technology.

Series C Financing

Nantero announced the successful completion of a $15 million Series C round of funding during the first quarter of 2005. Globespan Capital Partners was the lead investor in the round. Ullas Naik, a managing director with Globespan, based in its Boston office, was added to Nantero's board of directors. Returning institutional investors included Charles River Ventures, Draper Fisher Jurvetson, Harris & Harris Group, and Stata Venture Partners.

Post–Series C Funding Evolution

In spring 2006, Nantero announced another milestone with the fabrication and successful testing of a twenty-two-nanometer NRAM memory switch. This switch demonstrated that NRAM was scalable to numerous process technology nodes over several decades. Nantero's NRAM switches were tested by writing and reading data using three nanosecond cycle times. This gave NRAM the potential to match the fastest memories in production at the time. The switches were fabricated using the company's proprietary CNT fabric. The results demonstrated that NRAM could be the standalone and embedded memory of choice, combining the nonvolatility of flash with the speed of SRAM and the density of DRAM. The test results demonstrated that NRAM could be scaled for many future generations, with scaling expected to continue below the five-nanometer technology node. Nantero announced collaboration with ON Semiconductor in the spring of 2006 to jointly develop CNT technology, continuing ongoing efforts to integrate CNTs in CMOS fabrication.

In fall 2006, Nantero announced that it had resolved all major obstacles that had prevented CNTs from being used in mass production in semiconductor fabs. This was a major milestone for the company. Nanotubes were widely acknowledged to hold great promise for the future of semiconductors, but most experts had predicted it would be a decade or more before they would become a viable material. This belief was the result of several historic obstacles that had prevented their use, including a previous inability to position them reliably across entire silicon wafers as well as contamination that had made the nanotube material incompatible with semiconductor fabs.

Nantero developed a method for positioning CNTs reliably on a large scale by treating them like a fabric that can be deposited using methods like spin coating and then patterned using conventional lithography and etching, common to all CMOS processes in semiconductor fab. Additionally, Nantero developed a method for purifying CNTs to the standards required for use in a production semiconductor fab: this meant CNTs consistently containing less than twenty-five parts per billion of any metal contamination. With these innovations, Nantero became the first company in the world to introduce and use CNTs in mass-production semiconductor fabs.

In summer 2007, Nantero licensed biomedical sensors to Alpha Szenszor, a newly formed company also based in Woburn, Massachusetts. Alpha Szenszor was founded to develop a suite of sensor products in the medical segment, including portable and cost-effective detectors for infectious diseases like the human immunodeficiency virus (HIV). Alpha Szenszor's cofounder, Steve Lerner, is an industry veteran with two decades of product development and industrialization experience. Also during this time, Nantero announced it was working with Hewlett-Packard to explore the use of HP inkjet technology and the company's CNT formulation to create flexible electronics products and develop low-cost printable memory applications. Nantero used HP's thermal inkjet pico-fluidic system R&D tool to gauge the company's inkjet technology for printable memory applications that could be used in a wide range of applications, including low-cost radio-frequency identification (RFID) tags.

During summer 2008, Nantero announced collaboration with SVTC Technologies to accelerate the commercialization of CNT-based electronics products. The collaboration was part of SVTC's mission to enable the commercialization of new process and device developments in the semiconductor, microelectromechanical systems (MEMS), and related nanotechnology domains with support for a direct route between the work completed in SVTC's facilities to high-volume manufacturing. The collaboration between Nantero and SVTC offered CNT device development capabilities for customers targeting a wide range of applications, including photovoltaics (solar cells), sensors, MEMS, light-emitting diodes (LEDs), and other semiconductor-based devices.

In 2008, Nantero announced a major milestone that involved the acquisition of its government business unit by Lockheed Martin. At the time, Lockheed Martin was a leader in the R&D and application of nanotechnology to future military and intelligence applications. Lockheed Martin entered an exclusive license arrangement with Nantero for government-specific applications of the company's broad intellectual property portfolio. As part of the purchase of Nantero's government business unit, some thirty employees, including Nantero cofounder and COO Brent Segal, joined Lockheed Martin. Lockheed Martin's Advanced Technology Center, a unit of the Lockheed Martin Space Systems Company, would manage the Nantero unit in the future. Deal terms were undisclosed. The year 2008 also saw Nantero ranked fifty-fourth in *Inc. Magazine's* "Annual List of America's 500 Fastest-Growing Private Companies" and second in the computer and electronics category.

Lockheed Martin announced in late 2009 that it had successfully tested a radiation-resistant version of Nantero's NRAM devices on a NASA shuttle mission in May 2009 that involved servicing the Hubble Space Telescope. The experiment was a proof of concept that enabled the testing of launch and re-entry survivability, as well as basic functionality of the CNT memory in orbit throughout the shuttle mission. The mission represented a milestone in the development of NRAM technology for spaceflight applications.

Series D Financing Round

Nantero announced the closing of a Series D round of financing in 2013, which raised more than $15 million. The round included existing investors Charles River Ventures, Draper Fisher Jurvetson, Globespan Capital Partners, Harris & Harris Group, and Stata Venture Partners and also new strategic corporate investors. Among the notable new investors was Schlumberger, the leading supplier of technology, integrated project management, and information solutions to customers working in the oil and gas industry worldwide. Additionally, Michael Raam joined Nantero's advisory board. Ramm had held positions as CEO of SandForce, the successful solid-state drive (SSD) controller company and vice-president and general manager of the Flash Components Division of LSI after LSI acquired SandForce.

As 2009 was coming to a close, Nantero announced that Tsugio Makimoto, a Japanese semiconductor industry pioneer, had joined its advisory board. Makimoto had previously served as corporate advisor of the Sony Corporation in charge of semiconductor technology. Nantero noted that Makimoto's deep knowledge of and experience in the semiconductor industry, both in Japan and globally, would assist in bringing in new Asian customers.

Post–Series D Funding Evolution

During 2014, a team of researchers in Japan independently verified that Nantero's NRAM had excellent properties and could be used for applications ranging from main memory to storage. Leading the Japanese NRAM research team was Ken Takeuchi, professor in the Department of Electrical, Electronic, and Communication Engineering of the Faculty of Science and Engineering at Chuo University. Takeuchi had previously led Toshiba's NAND flash memory circuit design for fourteen years. The details of the NRAM technology were announced in a lecture titled "23 Percent Faster Program and 40 Percent Energy Reduction of Carbon Nanotube Non-volatile Memory with Over 10^{11} Endurance" (lecture number T11-3) at the Symposia on VLSI Technology and Circuits in Honolulu, Hawaii, in 2014.

Series E Financing Round

In spring 2015, Nantero announced the completion of a $31.5 million Series E capital raise. The round included new investors and participation from existing investors Charles River Ventures, Draper Fisher Jurvetson, Globespan Capital Partners, and Harris

& Harris Group. The additional funding was raised to help accelerate NRAM as the leading next-generation memory for both storage-class memory and as a replacement for flash and DRAM. In addition to the funding, Nantero added two new advisory board members: Stefan Lai, a former Intel senior executive who co-invented the EPROM tunnel oxide (ETOX) flash memory cell and led the company's phase change memory team, and Yaw Wen Hu, a former executive vice-president and current board member of Inotera Memories, where he oversaw new DRAM technology transfer and the development of wafer-level packaging. Hu had been executive vice-president and COO Officer for Silicon Storage Technology, where he was responsible for SuperFlash technology development, working with a team to develop a new memory cell concept into a high-volume product shipment and establish it as the choice of technology for embedded flash applications.

In summer 2015, Nantero announced that arrays of its new generation of super-fast, high-density memory (NRAM) were being independently tested by Chuo University. The results, which revealed excellent performance and reliability, would be presented in a technical paper at the 2015 International Conference on Solid State Devices and Materials. Nantero also announced that former TSMC executive Shang-Yi Chiang would be joining the company's advisory board. With more than forty years of experience in the semiconductor industry, including as co-COO and executive vice-president at TSMC, Chiang has contributed to the R&D of CMOS, NMOS, Bipolar, DMOS, SOS, SOI, GaAs lasers, LED, E-Beam lithography, and silicon solar cells. Additionally, Lee Cleveland was added to Nantero's executive management team as vice-president of design, with responsibility for leading Nantero's complete chip design team. Cleveland had previously been in charge of flash design at Spansion and AMD. Nantero also announced the appointment of renowned memory industry executive Ed Doller to its advisory board. Doller was previously vice-president and chief strategist of the NAND Solutions Group at

Micron, where he also served as vice-president and general manager of enterprise storage and vice-president and chief memory systems architect.

Summary

Nantero has spent the past fifteen years engaged in intensive R&D and commercialization of CNT-based memory devices. As a pioneer in nanotechnology, Nantero is the first company to develop semiconductor products using this material in the production of CMOS fabs. Nantero's NRAM technology has several characteristics that make it attractive as a next-generation technology for standalone and embedded uses, including the following:

- High endurance: Proven to operate for cycles of orders of magnitude more than flash
- Faster read-and-write: Same as DRAM, hundreds of times faster than NAND
- CMOS compatible: Works in standard CMOS fabs with no new equipment needed
- Limitless scalability: Designed to scale below five nanometers in the future
- High reliability: Will retain memory for more than 1,000 years at 85°C and more than ten years at 300°C
- Low Cost: Simple structure, can be 3D multilayer and multilevel cell (MLC)
- Low Power: Zero in standby mode, 160 times lower write energy per bit than NAND

One of the major advantages of Nantero's NRAM technology is the ability to consolidate DRAM with flash memory. NRAM has been shown to be both as fast as DRAM and nonvolatile, like flash. These characteristics are attractive to manufacturers seeking to create smaller, more powerful electronic devices in the future. Potential future applications of NRAM technology include instant-on laptops, next-generation enterprise systems, virtual screens, rolled-up tablets, 3D videophones, and other products needing large amounts of fast memory. Such applications target a wide range of markets, including consumer electronics, mobile computing, wearables, the Internet of Things, enterprise storage, automotive, government, military, and space.

Over the past fifteen years, Nantero has raised more than $87 million and generated more than $70 million in revenue. The company's nanotechnology is presently under development in multiple worldclass manufacturing facilities. Nantero has more than a dozen major corporate partners actively engaged in commercializing NRAM technology. Sample NRAMs have demonstrated production-ready yields (i.e., over 99.999 percent). Performance for both standalone and embedded memory applications are superior to anything else on the market. Nantero is currently sampling memory test chips to customers at a time when there is large demand for new high-density, standalone memory and high-reliability, scalable, embedded memory.

Nantero has the potential to scale its business significantly beyond NRAM through the licensing of its large CNT patent portfolio for other applications in a wide range of segments. The company's business model is similar to ARM Holdings, a peer that today is generating more than $0.5 billion in revenue per year. As chips scale further down into the quantum realm with advanced nanomaterials that have properties that extend beyond silicon, there is ample opportunity for Nantero and its CNT technology in the $330 billion–plus semiconductor industry.

From Nantero's case study, we see the time, effort, and patience required to commercialize a transformative deep science–enabled technology relative to a social media application circa 2012–2014. Nantero's path to commercialization extends over a period of fifteen years. Unlike many of its peers in semiconductors and

Financing Rounds and Selected Corporate Milestones

2001	2003	2004–2005	2006	2007–2011	2012–2015
Company founded	Series B Round: $10.5 m	Series C Round: $15 m	Resolved all major obstacles that had been preventing CNT from being used in mass production in semiconductor fabs	Sale of government business unit to Lockheed Martin	Series D Round: $15 m Series E Round: $31.5 m
Series A Round: $6 m	Large-scale array of nanotubes on silicon wafers	Seminal patent issued covering CNT films and fabrics	Fabricated and successfully tested a 22-nanometer NRAM memory switch that can scale down to 5 nanometers	A radiation-resistant version of NRAM developed jointly with Lockheed Martin tested on NASA space shuttle mission	Collaboration with HP semiconductor research verifies
	Collaboration with ASML	Total patent count: 10 granted and more than 40 pending	ON Semiconductor collaboration to jointly develop CNT Technology	Collaboration with Hewlett-Packard to create flexible CNT electronics products	NRAM has excellent properties as universal memory device that can be used for various applications ranging from main memory to storage
	Mohan Rao joins scientific advisory board	CNT development project with LSI Logic	Actively seeks European and Asian partners to further advance NRAM commercialization	Collaboration with SVTC Technologies to accelerate commercialization of nanotube-based electronic products	Tsugio Makimoto joins advisory board
		Joint evaluation of CNT-based electronics with BAE Systems		License rights for biomedical sensors to Alpha Szenszor	Stefan Lai, Yaw Wen Hu, and Ed Doller join advisory board
		$4.5 m award to develop CNT-based radiation-hard, non-volatile RAM		Ranked no. 54 in *Inc. Magazine*'s "Annual List of America's 500 Fastest-Growing Private Companies"	
		O. B. Bilous joins board of directors			

FIGURE AP2.1 The Nantero timeline. *Source:* Nantero

electronics, the company has progressed with the development of its innovative NRAM without accessing public capital markets through an initial public offering. Whether the company will be able to tap the public capital markets for growth capital in the future remains to be seen.

Nantero's evolution since the company's founding in 2001 is summarized in the accompanying figure ap2.1.

Notes

Introduction

1. Martin Bailey and Alok Chakrabarti, *Innovation and Productivity Crisis* (Washington, DC: The Brookings Institution, 1988), 119.

2. The term "economic dynamism" is used in this book to represent the underlying forces in the economy that drive growth in employment, income, output, productivity, and wealth creation. "Business dynamism," another term used in this book, is the process by which firms continually are born, fail, expand, and contract, as jobs are created, destroyed, and turned over.

3. Alan S. Blinder," The Mystery of Declining Productivity Growth," *Wall Street Journal*, May 14, 2015.

4. Robert Gordon, *The Rise and Fall of American Growth* (Princeton, NJ: Princeton University Press, 2016), 2.

5. Lucia Foster, Cheryl Grim, and John Haltiwanger, "Reallocation in the Great Recession: Cleansing or Not?," *Journal of Labor Economics* 34, no. S1 (January 2016): S293–S331.

6. Robert E. Litan and Ian Hathaway, "Declining Business Dynamism in the United States: A Look at States and Metros," *Brookings*, May 5, 2014, www.brookings.edu/research/declining-business-dynamism-in-the-united -states-a-look-at-states-and-metros/.

7. James Surowiecki, "Why Startups Are Struggling," *MIT Technology Review* 119, no. 4 (2016): 112–115.

8. Amy Bernstein, "The Great Decoupling: An Interview with Erik Brynjolfsson and Andrew McAfee," *Harvard Business Review*, June 2015.

9. George Gilder, *The Scandal of Money: Why Wall Street Recovers but the Economy Never Does* (Washington, DC: Regnery, 2016), 121.

1. Deep Science Disruption

1. Thomas L. Hankins, *Science and the Enlightenment* (New York: Cambridge University Press, 1985), 1.

2. Henry Adams, *The Education of Henry Adams* (Boston: Houghton Mifflin, 1918). Available at www.bartleby.com/159/25.html.

3. "Tesla Quotes by Dr. Charles F. Scott," Tesla Universe, https://teslauniverse .com/nikola-tesla/quotes/authors/dr-charles-f-scott.

4. Quote from Arthur Zajonc, *Catching the Light* (New York: Oxford University Press, 1993), 145.

5. We have chronicled the evolution of quantum physics elsewhere and refer interested readers to that material for a more comprehensive review of the development of this branch of deep science. See Stephen R. Waite, *Quantum Investing* (Mason, OH: Thomson Learning, 2002/2004).

6. Richard Feynman, *The Character of Physical Law* (New York: Random House, 1994), 129.

7. Robert J. Gordon, "Is U.S. Economic Growth Over? Faltering Innovation Confronts the Six Headwinds" (Center for Economic Policy Research, Policy Insight No. 63, September 2012).

8. United States Patent and Trademark Office, "The U.S. Patent System Celebrates 212 Years," press release #02-26, April 9, 2002.

9. Patent data are from the U.S. Patent and Trademark Office (www .uspto.gov/).

10. The Patent Act of 1836 established the Commissioner for Patents for the U.S. Patent and Trademark Office.

11. Elhanan Helpman, *General Purpose Technologies and Economic Growth* (Cambridge, MA: MIT Press, 1998).

12. Joseph A. Schumpeter, *Capitalism, Socialism and Democracy* (New York: Harper, 1942), 83.

13. Eric D. Beinhocker, *The Origin of Wealth* (Boston: Harvard Business School Press, 2007).

14. Ibid., 75.

15. George Gilder, *Knowledge and Power: The Information Theory of Capitalism and How It Is Revolutionizing Our World* (Washington, DC: Regnery, 2013).

16. David Warsh, *Knowledge and the Wealth of Nations* (New York: Norton, 2006).

17. For more on this subject, see Matt Ridley, *The Rational Optimist* (New York: Harper, 2010).

18. Jeremy Atack, "Long-Term Trends in Productivity," in *The State of Humanity*, ed. Julian Simon (Malden, MA: Blackwell, 1995), 169.

19. W. Bernard Carlson, *Tesla: Inventor of the Electrical Age* (Princeton, NJ: Princeton University Press, 2013).

20. Atack, "Long-Term Trends in Productivity."

21. Real per capital GDP is measured in constant 2009 dollars. Data from Louis Johnston and Samuel H. Williamson, "What Was the U.S. GDP Then?," MeasuringWorth, 2016, www.measuringworth.com/datasets/usgdp/result.php.

22. Max Tegmark and John Archibald Wheeler, "One Hundred Years of Quantum Mysteries," *Scientific American*, February 2001, 68–75.

2. The U.S. Deep Science Innovation Ecosystem

1. Subhra B. Saha and Bruce A. Weinberg, "Estimating the Indirect Economic Benefits of Science," November 15, 2010, www.nsf.gov/sbe/sosp/econ/weinberg.pdf; see also www.sciencecoalition.org/federal_investment. For an opposing view, see Colin MacIlwain, "Science Economics: What Science Is Really Worth," *Nature* 465 (2010): 682–684, www.nature.com/news/2010/100609/full/465682a.html.

2. Battelle and R&D Magazine, *2014 Global R&D Funding Forecast*, December 2013, http://www.rdmag.com/sites/rdmag.com/files/gff-2014-5_7%20875X10_0.pdf.

3. Vannevar Bush, *Science: The Endless Frontier; A Report to the President* (Washington, DC: Office of Scientific Research and Development, July 1945).

4. Judith Albers and Thomas R. Moebus, *Entrepreneurship in New York: The Mismatch Between Venture Capital and Academic R&D* (Geneseo, NY: Milne Library, SUNY Geneseo, 2013), 30.

5. Congressional Budget Office, *Updated Budget Projections: Fiscal Years 2013 to 2023*, May 14, 2013, https://www.cbo.gov/publication/44172.

6. MIT Committee to Evaluate the Innovation Deficit, *The Future Postponed: Why Declining Investment in Basic Research Threatens a U.S. Innovation Deficit*, April 2015, http://dc.mit.edu/innovation-deficit.

7. See, for example, Jon Gertner, *The Idea Factory: Bell Labs and the Great Age of American Innovation* (New York: Penguin, 2013).

8. Ashish Arora, Sharon Belenzon, and Andrea Patacconi, "Killing the Golden Goose? The Decline of Science in Corporate R&D" (NBER Working Paper No. 20902, National Bureau of Economic Research, Cambridge, MA, January 2015).

9. Steve Blank, "Lean Innovation Management—Making Corporate Innovation Work," *Steve Blank* (blog), June 26, 2015, http://steveblank.com/2015/06/26/lean-innovation-management-making-corporate-innovation-work/. See also Gary P. Pisano, "You Need an Innovation Strategy," *Harvard Business Review*, June 2015, https://hbr.org/2015/06/you-need-an-innovation-strategy.

10. Barry Jaruzelski, Volker Staack, and Brad Goehle, "The Global Innovation 1000: Proven Paths to Innovation Success," *Strategy + Business*, October 28, 2014, www.strategy-business.com/article/00295?gko=b91bb.

11. Ibid.

12. Josh Lerner, "Corporate Venturing," *Harvard Business Review*, October 2013, https://hbr.org/2013/10/corporate-venturing.

13. Randall Smith, "As More Companies Invest in Start-Ups, Soft Market Poses a Test," *New York Times*, April 19, 2016.

14. Alex Philippidis, "Despite Big Pharma Retreat, R&D Spending Advances," *GEN: Genetic Engineering & Biotechnology News*, March 15, 2015, http://www.genengnews.com/gen-articles/despite-big-pharma-retreat-rd-spending-advances/5446?q=Pharma%20spending%20$1.8%20billion.

15. Thomas J. Hwang, "Stock Market Returns and Clinical Trial Results of Investigational Compounds: An Event Study Analysis of Large Biopharmaceutical Companies," *PLOS ONE*, August 7, 2013, http://journals.plos.org/plosone/article?id=10.1371/journal.pone.0071966; Michael D. Hamilton, "Trends in Mid-stage Biotech Financing" (independent study, Tuck School of Business at Dartmouth, 2011), http://docplayer.net/10258764-Trends-in-mid-stage-biotech-financing.html.

16. Gary P. Pisano and Willy Shih, *Producing Prosperity: Why America Needs a Manufacturing Renaissance* (Boston: Harvard Business Review Press, 2012), 66.

17. Joseph Schumpeter, *Capitalism, Socialism and Democracy* (New York: Harper, 1942), chapter 7.

18. George Gilder, *Knowledge and Power: The Information Theory of Capitalism and How It Is Revolutionizing Our World* (Washington, DC: Regnery, 2013), 33.

19. Commitments to venture capital funds increased following clarification by the U.S. Department of Labor of the Employment Retirement Income Security Act's (ERISA's) Prudent Man Rule in 1979. Before then, the rule stated that pension managers had to invest with the care of a "prudent man." As a result, many pension funds avoided investing in venture capital entirely based on the belief that a fund's investment in a startup company could be viewed as imprudent. In early 1979, the Department of Labor ruled that portfolio diversification was a consideration in determining the prudence of an individual investment. Hence, the ruling implied that an allocation of part of a portfolio to venture funds would not be imprudent. This clarification opened the door for pension funds to invest in venture capital.

20. Josh Lerner, *The Architecture of Innovation* (Boston: Harvard Business Review Press, 2012).

3. Deep Science and the Evolution of American Venture Capital

1. Spencer E. Ante, *Creative Capital: Georges Doriot and the Birth of Venture Capital* (Boston: Harvard Business Publishing, 2008), 75.

2. Ibid.

3. Ibid., 111.

4. The National Science Foundation defines basic science as research that seeks to "gain more complete knowledge or understanding of the fundamental aspects of phenomena and of observable facts, without specific applications toward processes or products in mind." For more on this topic, see section 6 of the third annual report (1953) of the National Science Foundation, www.nsf.gov/pubs/1953/annualreports/ar_1953_sec6.pdf.

5. Gary W. Matkin, *Technology Transfer and the University* (New York: Macmillan, 1990), 9.

6. Yong S. Lee and Richard Gaertner, "Translating Academic Research to Technological Innovation," in *Technology Transfer and Public Policy*, ed. Yong S. Lee (Westport, CT: Quorum, 1997), 8.

7. "Vannevar Bush," *Wikipedia: The Free Encyclopedia*, https://en.wikipedia.org/w/index.php?title=Vannevar_Bush&oldid=742158064.

8. Ante, *Creative Capital*, 108.

9. Ibid., 112.

10. Ibid., 112.

11. Ibid., 173.

12. Ibid., 124.

13. Ibid., 233.

14. Cynthia Robbins-Roth, *From Alchemy to IPO: The Business of Biotechnology* (New York: Basic Books, 2001), 13.

15. Peter Thiel, *Zero to One: Notes on Startups, or How to Build the Future* (New York: Crown Business, 2014), 86.

16. Richard Smith, Robert Pedace, and Vijay Sathe, "VC Fund Financial Performance: The Relative Importance of IPO and M&A Exits for Venture Capital Fund Financial Performance," *Financial Management* 40, no. 4 (2011): 1031.

17. John Koetsier, "Halo Report: Average Angel Investment Up 23% as Web, Health, and Mobile Are 72% of All Deals," *VentureBeat*, July 19, 2013, http://venturebeat.com/2013/07/19/halo-report-average-angel-investment -up-23-as-web-health-and-mobile-are-72-of-all-deals/.

18. Policymakers and the media often cite R&D spending figures and their associated annual and historical growth rates as a way of gauging the pace of innovation within a nation or nationally from a global perspective. As Steve Jobs once pointed out, there is far more to innovation than money spent on research and development. Jobs noted that when Apple came up with the innovation of the Mac personal computer—a truly revolutionary device—IBM was spending at least one hundred times more on R&D.

19. See the November 9, 1998, issue of *Fortune* magazine.

4. Diversity Breakdown in Venture Investing

1. Judith Albers and Thomas R. Moebus, *Entrepreneurship in New York: The Mismatch Between Venture Capital and Academic R&D* (Geneseo, NY: Milne Library, SUNY Geneseo, 2013), 23.

2. The media and entertainment segment is included with software owing to the ongoing process of digitization. Additionally, this segment is highly social by nature and falls within the social media sector—a segment of intense focus for U.S. venture capital investors currently.

3. All U.S. venture capital data are from PricewaterhouseCoopers (PwC)/ National Venture Capital Association (NVCA) MoneyTree.

4. Deep science venture deals are classified as deals occurring in the following eight PwC/NVCA segments: biotech, medical devices, computer and peripherals, semiconductors, electronics and instruments, networking equipment, telecommunications, and industrial/energy.

5. Per Bak, *How Nature Works: The Science of Self-Organized Criticality* (New York: Springer-Verlag, 1996), 1.

6. Michael J. Mauboussin, *More Than You Know: Finding Financial Wisdom in Unconventional Places* (New York: Columbia University Press, 2008), 197.

7. Charles Mackay, *Extraordinary Popular Delusions and the Madness of Crowds* (1841; repr., New York: Wiley Investment Classics, 1996); Charles P. Kindleberger, *Manias, Panics, and Crashes: A History of Financial Crises*, 3rd ed. (New York: Wiley Investment Classics, 1996).

8. Steve Blank, "Why Facebook Is Killing Silicon Valley," *Steve Blank* (blog), May 21, 2012, http://steveblank.com/2012/05/21/why-facebook-is-killing-silicon-valley/.

9. Ibid.

10. Digi-Capital, "Average Mobile Unicorn Now Worth Over $9 Billion," *Digi-Capital* (blog), August 2015, http://www.digi-capital.com/news/2015/08/average-mobile-unicorn-now-worth-over-9-billion/#.WD7mUKIrLdR.

11. Fred Wilson, "What Can It Be Worth?," *AVC* (blog), May 26, 2015, http://avc.com/2015/05/what-can-it-be-worth/?utm_source=feedburner&utm_medium=feed&utm_campaign=Feed%3A+AVc+%28A+VC%29.

12. Michael J. Mauboussin, *The Success Equation: Untangling Skill and Luck in Business, Sports, and Investing* (Boston: Harvard Business School Publishing, 2012), 23–26.

13. Charles Duhigg, *Smarter Faster Better: The Transformative Power of Real Productivity* (New York: Random House, 2016), 167–204.

14. Blank, "Why Facebook Is Killing Silicon Valley."

15. Mark Suster, "Understanding Changes in the Software and Venture Capital Industries," *Both Sides of the Table*, June 28, 2011, www.bothsidesofthetable.com/2011/06/28/understanding-changes-in-the-software-venture-capital-industries/.

16. Blank, "Why Facebook Is Killing Silicon Valley."

17. Tyler Durden, "SEC Goes Unicorn Hunting: Regulator to Scrutinize How Funds Value Tech Startups," *ZeroHedge*, November 18, 2015, www.zerohedge.com/news/2015-11-18/sec-goes-unicorn-hunting-regulator-scrutinize-how-funds-value-tech-startups.

18. Aileen Lee, "Welcome to the Unicorn Club, 2015: Learning from Billion-Dollar Companies," *TechCrunch*, July 18, 2015, http://techcrunch

.com/2015/07/18/welcome-to-the-unicorn-club-2015-learning-from-billion
-dollar-companies/#.kbzhtz:Lhpo.

19. Richard Thaler, Amos Tversky, Daniel Kahneman, and Alan Schwartz, "The Effect of Myopia and Loss Aversion on Risk Taking: An Experimental Test," *Quarterly Journal of Economics* 112, no. 2 (1997): 647–661.

5. Fostering Diversity in Venture Investing

1. Victor W. Hwang and Greg Horowitt, *The Rainforest: The Secret to Building the Next Silicon Valley* (Los Altos Hills, CA: Regenwald, 2012).

2. David Weild and Edward Kim, *Why Are IPOs in the ICU?* (Chicago: Grant Thornton, November 28, 2008), www.grantthornton.com/staticfiles /GTCom/files/GT%20Thinking/IPO%20white%20paper/Why%20are%20 IPOs%20in%20the%20ICU_11_19.pdf, 3.

3. Ibid., 7.

4. See the National Venture Capital Association website: http://nvca.org.

5. Weild and Kim, *Why Are IPOs in the ICU?*, 4.

6. David Weild, Edward Kim, and Lisa Newport, "Making Stock Markets Work to Support Economic Growth: Implications for Governments, Regulators, Stock Exchanges, Corporate Issuers and Their Investors" (OECD Corporate Governance Working Papers, no. 10, Paris, OECD Publishing, 2013).

7. Ibid.

8. Weild and Kim, *Why Are IPOs in the ICU?*

9. Paul M. Healy, "How Did Regulation Fair Disclosure Affect the U.S. Capital Market? A Review of the Evidence," Harvard Business School, December 6, 2007, www.frbatlanta.org/-/media/Documents/news/conferenc es/2008/08FMC/08FMChealy.pdf.

10. Laura J. Keller, Dakin Campbell, Alastair Marsh, and Stephen Morris, "An Inside Look at Wall Street's Secret Client List," *Bloomberg*, March 24, 2016, http://www.bloomberg.com/news/articles/2016-03-24/wall -street-s-0-01-an-inside-look-at-citi-s-secret-client-list.

11. Peter Coy, "IPOs Get Bigger But Leave Less for Public Investors," *Bloomberg*, July 24, 2014, www.bloomberg.com/news/articles/2014-07-24/ipos-get -bigger-but-leave-less-for-public-investors.

12. Jason Voss, "The Decline in Stock Listings Is Worse than You Think," *CFA Institute: Enterprising Investor* (blog), September 30, 2013, https://blogs.cfainstitute.org/investor/2013/09/30/the-decline-in-stock- listings-is-worse-than-you-think/.

13. IssueWorks Inc., 2013–2014, as presented to Harris & Harris Group, July 8, 2014.

14. Weild, Kim, and Newport, "Making Stock Markets Work to Support Economic Growth."

15. Hatim Tyabji and Vijay Sathe, "Venture Capital Firms in America: Their Caste System and Other Secrets," *Ivey Business Journal*, July/August 2010, http://iveybusinessjournal.com/publication/venture-capital-firms-in-america-their-caste-system-and-other-secrets/.

16. Luke Timmerman, "Who's Still Active Among the Early-Stage Biotech VCs?" *Exome*, July 2, 2012, www.xconomy.com/national/2012/07/02/whos-still-active-among-the-early-stage-biotech-vcs/.

17. Ibid.

18. "Michael Dell and Silver Lake Complete Acquisition of Dell," *Dell*, October 29, 2013, www.dell.com/learn/us/en/uscorp1/secure/acq-dell-silverlake.

19. Michael J. Mauboussin and Dan Callahan, "A Long Look at Short-Termism: Questioning the Premise," *Journal of Applied Corporate Finance* 27, no. 3 (Summer 2015): 70–82.

20. George Gilder, *The Scandal of Money: Why Wall Street Recovers but the Economy Never Does* (Washington, DC: Regnery, 2016).

21. Charles Murray, *By the People: Rebuilding Liberty Without Permission* (New York: Crown Forum, 2015).

22. Chance Barnett, "Trends Show Crowdfunding to Surpass VC in 2016," *Forbes*, June 9, 2015, www.forbes.com/sites/chancebarnett/2015/06/09/trends-show-crowdfunding-to-surpass-vc-in-2016/.

23. S. H. Salman, "The Global Crowdfunding Industry Raised $34.4 Billion in 2015, and Could Surpass VC in 2016," *DazeInfo*, January 12, 2016, https://dazeinfo.com/2016/01/12/crowdfunding-industry-34-4-billion-surpass-vc-2016/.

24. U.S. Securities and Exchange Commission, "SEC Adopts Rules to Facilitate Smaller Companies' Access to Capital," press release, March 25, 2015, www.sec.gov/news/pressrelease/2015-49.html.

25. Matthew Herper, "Why One Cancer Company Has Raised $300 Million in 12 Months Without an IPO," *Forbes*, August 5, 2014, www.forbes.com/sites/matthewherper/2014/08/05/why-this-cancer-fighting-company-has-raised-300-million-in-just-12-months/#2ea5125f69d4.

26. Katherine Tweed, "Bill Gates and Tech Billionaires Launch Clean Energy Coalition," *IEEE Spectrum*, December 3, 2015, http://spectrum.ieee.org/energywise/energy/renewables/bill-gates-and-tech-billionaires-launch-clean-energy-coalition.

6. Deep Science Venture Investing

1. Raymond Kurzweil, *The Age of Intelligent Machines* (Cambridge, MA: MIT Press, 1990), 8.

2. Louis-Vincent Gave, "Viva La Robolution?," GK Research, October 24, 2013.

3. Sam Shead, "Eric Schmidt: Advances in AI Will Make Every Human Better," *Tech Insider* (blog), March 8, 2016, http://www.businessinsider.com /eric-schmidt-advances-in-ai-will-make-every-human-better-2016-3.

4. Michael S. Malone, *The Microprocessor: A Biography* (New York: Springer-Verlag, 1995), 251.

5. Freeman Dyson, "Our Biotech Future," *New York Review of Books*, July 19, 2007, http://www.nybooks.com/articles/2007/07/19/our-biotech -future/.

6. This statement was made during a talk at a 1977 meeting of the World Future Society in Boston. See, however, the context surrounding it at Snopes. com, *Fact Check: Home Computing*, "Did Digital founder Ken Olsen say there was 'no reason for any individual to have a computer in his home?'" at http://www.snopes.com/quotes/kenolsen.asp.

7. Chris Anderson, *Makers: The New Industrial Revolution* (New York: Crown Business, 2012), 15.

7. Our Choice Ahead

1. Ari Levy, "Tech IPO Market Shows Promise After Dead First Half of 2016," CNBC.com, http://www.cnbc.com/2016/09/09/tech-ipo-market -shows-promise-after-dead-first-half-of-2016.html.

2. George Gilder, *The Scandal of Money: Why Wall Street Recovers but the Economy Never Does* (Washington, DC: Regnery, 2016), 128.

3. Scott Grannis, "Household Wealth Increases, Leverage Declines," *Calafia Beach Pundit* (blog), June 11, 2015, http://scottgrannis.blogspot. com/2015/06/household-wealth-increases-leverage.html.

4. Gordon E. Moore, "Cramming More Components Onto Integrated Circuits," *Electronics* 38, no. 8 (April 1965): 114–17, http://web.eng.fiu.edu /npala/eee6397ex/gordon_moore_1965_article.pdf.

5. W. Michael Cox and Richard Alm, *Myths of Rich and Poor: Why We're Better Off than We Think* (New York: Basic Books, 1999), 116.

6. Richard Foster and Sarah Kaplan, *Creative Destruction: Why Companies That Are Built to Last Underperform the Market—and How to Successfully Transform Them* (New York: Currency Doubleday, 2001), 294.

7. Michael Dell, "Wanted: 600 Million New Jobs," *LinkedIn Pulse*, June 27, 2015, www.linkedin.com/pulse/wanted-600-million-new-jobs-michael-dell.

8. It is noteworthy that the percentage of bachelor's degrees conferred in the fields of science, technology, engineering, and mathematics in the United States has fallen below 25 percent in recent years. See "Bachelor's Degrees Conferred in Science, Technology, Engineering, and Mathematics Fields," *United States Education Dashboard*, http://dashboard.ed.gov/moreinfo.aspx?i=m&id=5&wt=0.

9. For more information, visit the USA Science and Engineering Festival website: www.usasciencefestival.org.

10. Jonathan Rothwell, "The Hidden STEM Economy," *Brookings*, June 10, 2013, www.brookings.edu/research/reports/2013/06/10-stem-economy-rothwell.

11. National Science Board, *Revisiting the STEM Workforce* (Arlington, VA: National Science Board, February 4, 2015).

12. Cromwell Schubarth, "Intel Capital, Silicon Valley Bank Chiefs See Signs of a Bubble," *Silicon Valley Business Journal*, April 13, 2015, www.bizjournals.com/sanjose/blog/techflash/2015/04/intel-capital-silicon-valley-bank-chiefs-see-signs.html.

13. "Open Letter on the Digital Economy," http://openletteronthedigitaleconomy.org.

14. Yuliya Chernova, "Early Tesla Motors Investors Raise $400 Million Impact VC Fund," *Wall Street Journal*, June 23, 2015, http://blogs.wsj.com/venturecapital/2015/06/23/early-tesla-motors-investors-raise-400-million-impact-vc-fund/.

15. Cited in Victor W. Hwang and Greg Horowitt, *Rainforest: The Secret to Building the Next Silicon Valley* (Los Altos Hills, CA: Regenwald, 2012), 130.

Appendix 1. The Case of D-Wave Systems

1. Richard P. Feynman, "Simulating Physics with Computers," *International Journal of Theoretical Physics* 21, no.6–7 (1982): 467–488; Richard P. Feynman, "Quantum Mechanical Computers," *Optics News* 11, (1985): 11–20.

2. David Deutsch, "Quantum Theory, the Church–Turing Principle, and the Universal Quantum Computer," *Proceedings of the Royal Society London* A, no. 400 (1985): 97–117.

3. Peter W. Shor, "Algorithms for Quantum Computation: Discrete Logarithm and Factoring," *Proceedings of the 35th Annual Symposium on Foundations of Computer Science* (1994): 124–134; Peter W. Shor, "Polynomial-Time Algorithms for Prime Factorization and Discrete Logarithms on a Quantum Computer," *SIAM Journal on Computing* 26, no. 5 (1997): 1484–1509.

4. Peter W. Shor, "Scheme for Reducing Decoherence in Quantum Computer Memory," *Physical Review A* 52, no. 4 (1995): R2493–R2496.

5. Lov Grover, "A Fast Quantum Mechanical Algorithm for Database Search," *Proceedings of the 28th Annual ACM Symposium on the Theory of Computing* (May 1996): 212.

6. Nicolas J. Cerf, Lov K. Grover, and Colin P. Williams, "Nested Quantum Search and Structured Problems," *Physical Review A* 61, no. 3 (2000): 032303.

7. Daniel S. Abrams and Seth Lloyd, "Quantum Algorithm Providing Exponential Speed Increase for Finding Eigenvalues and Eigenvectors," *Physical Review Letters* 83, no. 24 (1999): 5162–5165.

8. For a more in-depth discussion of quantum computing, see George Johnson, *A Shortcut Through Time: The Path to the Quantum Computer* (New York: Knopf, 2003); see also Seth Lloyd, *Programming the Universe: A Quantum Computer Scientist Takes on the Cosmos* (New York: Knopf, 2006).

9. Alan D. MacCormack, Ajay Agrawal, and Rebecca Henderson, "D-Wave Systems: Building a Quantum Computer," Harvard Business School Case 604-073 (Boston: Harvard Business School Publishing, April 26, 2004).

10. "Quantum Computing Firm D-Wave Systems Announces Milestone of 100 U.S. Patents Granted, Patent Portfolio also Rated #4 in Computing Systems by IEEE Spectrum in Latest Quality Assessment," *D-Wave Systems*, June 20, 2013, http://www.dwavesys.com/updates/quantum-computing-firm-d-wave-systems-announces-milestone-100-us-patents-granted-patent.

11. Hartmut Neven, Vasil S. Denchev, Geordie Rose, and William G. Macready, "Training a Binary Classifier with the Quantum Adiabatic Algorithm," *arxiv.org*, November 4, 2008, https://arxiv.org/abs/0811.0416.

12. Hartmut Neven, Vasil S. Denchev, Geordie Rose, and William G. Macready, "Training a Large Scale Classifier with the Quantum Adiabatic Algorithm," *arxiv.org*, December 4, 2009, https://arxiv.org/abs/0912.0779.

13. "Binary Classification Using Hardware Implementation of Quantum Annealing" (demonstration, Conference on Neural Information Processing Systems, Vancouver, British Columbia, December 7, 2009).

14. M. W. Johnson, et al., "Quantum Annealing with Manufactured Spins," *Nature* 473, no. 7346 (2011): 194–198.

15. Vadim N. Smelyanskiy, et al., "A Near-Term Quantum Computing Approach for Hard Computational Problems in Space Exploration," *arxiv .org*, April 12, 2012, https://arxiv.org/abs/1204.2821.

16. Immanuel Trummer and Christoph Koch, "Multiple Query Optimization on the D-Wave 2X Adiabatic Quantum Computer," *arxiv.org*, October 21, 2015, https://arxiv.org/abs/1510.06437.

17. N. G. Dickson, et al., "Thermally Assisted Quantum Annealing of a 16-Qubit Problem," *Nature Communications* 4, no. 1903 (2013): doi:10.1038/ncomms2920.

18. Tameem Albash and Daniel A. Lidar, "Decoherence in Adiabatic Quantum Computation," *Physical Review A* 91, no. 6 (2015): 062320.

19. "Interactive: Patent Power 2012," *IEEE Spectrum*, December 3, 2012, http://spectrum.ieee.org/static/interactive-patent-power-2012#anchor _comp_syst.

20. Lev Grossman, "The Quantum Quest for a Revolutionary Computer, *Time*, February 6, 2014, http://time.com/4802/quantum-leap/.

21. "50 Smartest Companies," *MIT Technology Review*, February 18, 2014, www2.technologyreview.com/tr50/2014/.

22. Tameem Albash, Walter Vinci, Anurag Mishra, Paul A. Warburton, and Daniel A. Lidar, "Consistency Tests of Classical and Quantum Models for a Quantum Annealer," *Physical Review A* 91, no. 4 (2015): 042314.

23. Seung Woo Shin, Graeme Smith, John A. Smolin, and Umesh Vazirani, "How 'Quantum' Is the D-Wave Machine?" *arxiv.org*, January 28, 2014 (last revised May 2, 2014), https://arxiv.org/abs/1401.7087.

24. Seung Woo Shin, Graeme Smith, John A. Smolin, and Umesh Vazirani, "Comment on 'Distinguishing Classical and Quantum Models for the D-Wave Device,'" *arxiv.org*, April 25, 2014 (last revised April 28, 2014), https://arxiv.org/abs/1404.6499.

25. T. Lanting, et al., "Entanglement in a Quantum Annealing Processor," *Physical Review X* 4, no. 2 (2014): 021041.

26. Sergio Boixo, et al., "Computational Role of Collective Tunneling in a Quantum Annealer," *arxiv.org*, November 14, 2014 (last revised February 19, 2015), https://arxiv.org/abs/1411.4036.

27. Kristen L. Pudenz, Tameem Albash, and Daniel A. Lidar, "Error Corrected Quantum Annealing with Hundreds of Qubits," *Nature Communications* 5, no. 3243 (2014): doi:10.1038/ncomms4243.

28. Walter Vinci, Tameem Albash, Gerardo Paz-Silva, Itay Hen, and Daniel A. Lidar, "Quantum Annealing Correction with Minor Embedding," *arxiv.org*, July 9, 2015, https://arxiv.org/abs/1507.02658.

29. Gili Rosenberg, Poya Haghnegahdar, Phil Goddard, Peter Carr, Kesh-eng Wu, and Marcos López de Prado, "Solving the Optimal Trading Trajectory Problem Using a Quantum Annealer," *arxiv.org*, August 22, 2015 (last revised August 11, 2016), http://arxiv.org/abs/1508.06182.

30. Helmut G. Katzgraber, Firas Hamze, and Ruben S. Andrist, "Glassy Chimeras Could Be Blind to Quantum Speedup: Designing Better Bench-marks for Quantum Annealing Machines," *Physical Review X* 4, no. 2 (2014): 021008; Martin Weigel, Helmut G. Katzgraber, Jonathan Machta, Firas Hamze, and Ruben S. Andrist, "Erratum: Glassy Chimeras Could Be Blind to Quantum Speedup: Designing Better Benchmarks for Quantum Annealing Machines," *Physical Review X* 5, no. 1 (2015): 019901.

31. Andrew D. King, Trevor Lanting, and Richard Harris, "Performance of a Quantum Annealer on Range-Limited Constraint Satisfaction Prob-lems," *arxiv.org*, February 7, 2015 (last revised September 3, 2015), https://arxiv.org/abs/1502.02098.

32. James King, Sheir Yarkoni, Mayssam M. Nevisi, Jeremy P. Hilton, and Catherine C. McGeoch, "Benchmarking a Quantum Annealing Processor with the Time-to-Target Metric," *arxiv.org*, August 20, 2015, https://arxiv.org/abs/1508.05087.

Appendix 2. The Case of Nantero

1. Dean Takahashi, "Intel's Gordon Moore Speculates on the Future of Tech and the End of Moore's Law," *VentureBeat*, May 11, 2015.

Index

Page numbers in italics indicate figures or tables.

academia, R & D and, 53, *55*, *56*
active firms, 153
Adams, Henry, 24–25
Adams Station, 26
additive manufacturing, 181–182, *182*
adiabatic model, of quantum computing, 214–215, 218
administrative law, 159
Advanced Research Project Agency, DOD, 86, 106, 215
Affordable Care Act, 158
Age of Electrification, 27, 31
aging, 57
agriculture, 192
Airbnb, 121
Alphabet, 173
Alpha Szenszor, 241
alternating-current machines, 43

alternative trading system (ATS), *151*
Amazon, 128, 172–173
American Competitiveness Initiative, 197–198
American Research and Development (ARD), 2, 80, 83–88, 92, 95, 141
Ames Research Center, 222, 226
Amgen Inc., 138, *139*, 141
Ampex, 94–95
ancient civilizations, 20
Anderson, Chris, 181
Anderson, Harlan, 2
Andreessen, Marc, 109
Android, 126
angel funds, 103–105, *104*, 128, 133
angel investors, 50, *104*, 125, 140, 141

Ante, Spencer, 80, 82
antibiotics, 57
APIs. *See* application program
   interfaces
Apple, 9, 13, 95, 105, 181, 202
application program interfaces
   (APIs), 216, 230
apps, 128
architectural venture capital, 93, 94
ARD. *See* American Research and
   Development
ARM Holdings, 235, 246
artificial intelligence (AI), 172, 191,
   202, 211
ASML, Nantero and, 237–238
asset classes, 100, *101*, 102
astronomy, 19, 20–21, 22, 23
Atari, 75, 91, 94–95
atoms, 28, 181–184
ATP, 106
ATS. *See* alternative trading system
automation, 173

Babbage, Charles, 21
bacteria, 57, 178
BAE Systems, 238
Baird Associates, 85
banks, 146, 147. *See also*
   investment banks
Bayesian psychology, 123–125
Beckman Instruments, 90
Beinhocker, Eric, 37
Bell, Alexander Graham, 58
Bell Labs, 1, 206
Bezos, Jeff, 66, 167, 222
Bilous, O. B., 238
binding, 43
biochemistry, 179
biological systems, 37
biology, 20
BIOLOGY+, 177–181

biotechnology, 16, 67, 93–94, 102,
   112–114, *113*, 115, 130, 168
bits, 181–184, 207–209
Blacksall, Frederick, 83
"black swan" events, 116
Blake, Bill, 220, 225
Blank, Julius, 90, 126
Blank, Steve, 119
Blodgett, Katharine, 59
Bock, Larry, 195
Boltzmann, Ludwig, 29
Boston Dynamics, 173
Boyer, Herbert, 96
brain chemistry, 57
Branson, Richard, 167
Breakthrough Energy Coalition,
   167
Bringing Down the House
   (Mezrich), 124–125
Brownell, Vern, 219, 228
Brush's dynamo, 25
Bush, George W., 197–198
Bush, Vannevar, 1, 53, 82–83, 89,
   90, 197, 201–202
business: creativity and, 41;
   development, stock market and,
   144–145; dynamism, 9, *10–11*,
   188; innovation and, 60–61;
   models, 15–17, 92, 93, 111, 130;
   start-up, 74, 77, 100, 119, 122,
   164, 194–195, 198–199. *See also*
   corporations
By the People (Murray), 157

California Institute of Technology,
   89, 215
capital, venture funds and, 99. *See
   also* venture capital
capitalism, 19, 36, 73
*Capitalism, Socialism and
   Democracy* (Schumpeter), 73

capital markets, 12, 147, 194, 203
carbon nanotubes (CNTs), 176,
    183–184, 234–236, 238–240,
    242, 245
causality principle, 29
celestial mechanics, 29
cells, 179
Cetus, 96
chaos theory, 29, 155
chemistry, 18, 20, 23, 59
chip technology, 15–16
Christensen, Clayton, 169
Churchill, Winston, 18, 19, 28
"churn," 192–193
Circo Products, 85
Citibank, 95, 125
Civil War, 9
classical mechanics, 23, 29,
    45, 170
classical physics, 189
clean energy, 66, 94, 98, 167
Cleveland, Lee, 243
climate, 58
clinical trials, 68
cloud, 127
CMOS. *See* complementary
    metal–oxide–semiconductor
CNTs. *See* carbon nanotubes
Cobb–Douglas production function,
    5
Cold War, 1
commercialization, 105; of clean
    energy, 167; deep science
    and, 18, 48, 51, 53, 63–72,
    64, 130–131, 135, 184;
    diversity breakdown and,
    162; economic dynamism and,
    31; entrepreneurs and, 167;
    innovation and, 108, 203; IP
    and, 31–32; of nanotechnology,
    187; at Nantero, 244, 246–247;

of NRAM technology, 239;
    offshore, 202; process, 154; of
    technology, 74, 115, 181
commissions, 146–147, 152
commodities, 37, 194
communication technologies, 34
companies: consumer-oriented, 130;
    development of, 166; Fortune
    500, 141; insurance, 84; Internet,
    110–111; listed, 145; public,
    78, 152; R & D payoff and, 14;
    venture capital and, 3, 3–4, 13,
    77–78
competition, commodities and, 194
complementary metal–oxide–
    semiconductor (CMOS), 221,
    237–238, 240, 245
complex adaptive behavior, 37
complexity, 21, 29, 37, 116, 118,
    190
complex systems, 116
Compton, Karl Taylor, 81, 83
computer programming, 21
computers, 1, 5, 39, 72, 102, 114,
    157, 168, 174, 190, 233. *See also*
    quantum computing
computer science, 14–15, 21
computer vision, 211
consolidation, 138, 139
Constitution, U.S., 32, 188–189
Consumer Protection Act, 157, 158
consumers, 26, 32, 37, 56, 130,
    192–193
consumption patterns, 35
Copernicus, Nicolaus, 29
Corning, Warren, 88
Corning Glass Works, 88
corporations, 51–52, 58–63, 66
cotton thread, 42, 43
co-tunneling, 205, 207
Cowboy Ventures, 130

Creative Capital (Ante), 80, 82
creative commerce technologies, 108
creative destruction, 31, 170, 171, 191, 193–194
Creative Destruction (Foster & Kaplan), 193–194
creative disruption, 188–194
creativity, 37, 41, 74
crowdfunding, 163–165
customer adoption, 136

d'Arbeloff, Alex, 87
DARPA. *See* Defense Advanced Research Projects Agency
Davis, Tommy, 91
Davis and Rock, 88, 91
DEC. *See* Digital Equipment Corporation
decimalization, 145, 146
deep science, *136*; advances in, 40, 169; business models, 15–16; commercialization and, 18, 48, 51, 53, 63–72, 64, 130–131, 135, 184; corporations and, 58–63; definition of, 18; disciplines of, 20–21; diversity and, 200; economic dynamism and, 40–49, 115, 183, 189; economy and, 36–41; entrepreneurship and, 36; evolution of, 17, 29, 49; incentive for, 144; Industrial Revolution and, 41, 170; innovation and, 17, 72, 109, 133, 137, 166–167, 174–176, 201, 202; innovation ecosystem, 52, 147, 160, 162–163, 203; investing, 103; IP and, 31–32; IPOs and, 185–186; knowledge and, 30; living standards and, 44–45; migration away from, 129;

product development from, 18; progression of, 170; R & D and, 75, 111; R & D payoff, 186–187; Silicon Valley and, 129; software and, 14–17, 200; technology and, 18–19, 30, 31, 33–34, 36–37, 39, 40, 48, 65, 67, 131, 140, 189, 192; timing of advances in, 38; venture capital and, 106, 159–160, 169–171, 175, 202
"Deep Science Innovation Ecosystem," 49
Defense Advanced Research Projects Agency (DARPA), 106, 215
Dell, Michael, 154, 194–195
Dell computers, 154–155
Dennis, Reid, 91
Department of Defense (DOD), 86, 106, 215
Descartes, René, 29
Deutsch, David, 206
development cycle, 104, 135
DeWolf, Nick, 87
diffusion rate, of new technology, *156*, 157
Digi-Capital, 121
Digital Equipment Corporation (DEC), 1–2, 87, 89, 181
digital processors, 207
digital revolution, 199
digital technologies, 12, 98, 102, 187, 199
disease, 171, 177–178, 179
disruption, creative, 188–194
disruptive innovation, 73, 169
disruptive venture capital, *93*
distribution, 128
diversity: complexity and, 118; deep science and, 200; innovation and, 71–72; investors, 116–117; R & D payoff and, 132

diversity breakdown, 7, 13, 103,
106–108; commercialization
and, 162; of early 2000s, 118;
in financial markets, 117, 147;
innovation and, 162; process,
202; R & D and, 187; software
and, 118, 122; in stock market,
118; venture capital and, 187,
200; in venture investing,
115–126, 129–130
DNA, 178, 180
DNA computing, 176
DNA-SEQ, 225
DOD. *See* Department of Defense
Dodd-Frank Wall Street Reform,
157, 158
Doller, Ed, 243–244
"Do Real-Output and Real-Wage
Measures Capture Reality? The
History of Lighting Suggests
Not" (Nordhaus), 39
Doriot, Georges, 81, 83–85, 88, 90
dot-com boom, 78
dot-com collapse, 98, 100, 142, 185
DRAM. *See* dynamic random-access
memory
Draper, Gaither and Anderson, 88
Draper, William H., III ("Bill"), 88,
91
Draper and Johnson Investment
Company, 88
drones, 172
Drucker, Peter, 202
drugs, 16, 66, 68–69, 71, 177
Duhigg, Charles, 123–124
DuPont, 58
D-Wave Systems, 175, 183, 205–
206; APIs, 216, 230; commercial
sales of, 219–220; D-Wave
2X, 226; D-Wave One system,
220; D-Wave Two system, 220,

222, 229; employees of, 228;
financing rounds, 213–214;
founding of, 212–213, 232;
Google and, 218–219, 222,
224, 226; headquarters, 213;
history of, 224; infrastructure
of, 228; IP strategy, 216, 223;
machine learning, 225; in
media, 223–224; NASA and,
222, 226; as nondeterministic,
209; patents of, 216, 223, 227;
processor, 221, 223, 224, 226,
228, 230; prototypes, 216;
publications of, 227; quantum
annealer, 205, 221, 223, 225,
227, 229; quantum effects in,
207–211; quantum processors
of, 216; QUBO problems, 216;
Rainier chip, 220; R & D, 230;
scalability of, 229, 232; Series A
financing round, 214; Series B
financing rounds, 215–217;
Series C financing round,
217–218; Series E financing
round, 220–221; Series F
financing round, 221–225; Series
G financing round, 225–227;
timeline, 231; 2X, 216, 219, 229;
USRA and, 226; "Vesuvius" chip,
220; visualization of workings, 210
dynamic random-access memory
(DRAM), 235, 244–245
dynamo, 24, 25, 34, 39, 58
Dyson, Freeman, 177

ecology, 20
e-commerce business models, 130
economic dynamism, 7, 108;
commercialization and, 31;
decline of, 19, 199; deep science
and, 40–49, 115, 183, 189;

economic dynamism (*continued*)
definition of, 188; entrepreneurs
and, 155; GPTs and, 35; growth
of, 51; of Industrial Revolution,
22; innovation and, 165, 168,
191–192, 196; living standards
and, 47, 190, 191; R & D
payoff and, 105; reduction of,
12; regulation and, 158; social
media and, 129; STEM and, 196;
technology and, 30, 33, 47, 183;
venture capital and, 12–13, 14,
51, 75, 79, 92, 116, 201; venture
investing and, 204; waning, 9
economic progress, 9
economic resources, 56
economics, 135
economic stagnation, 105
economic theory, 37, 134, *134*
economy: deep science and, 36–41;
GPTs and, *18*, 34, 35; Industrial
Revolution and, 170; innovation
and, 140, 204; modern, 189;
postwar, 4–5; quantum, 46;
R & D and, 5, 108; regulation
of, 81; technology and, 35;
venture capital and, 14, 72–79,
76; zero-sum, 186, 193
Edison, Thomas, 58
"Educate to Innovate" campaign,
197–198
Education of Henry Adams, The
(Adams), 24
"Effect of Myopia and Loss
Aversion on Risk Taking,
The: An Experimental Test"
(Thaler, Tversky, Kahneman, &
Schwartz), 131
Einstein, Albert, 29, 37, 190, 209
Eisenhower, Dwight, 86

elasticity, in R & D, 5
electricity, 19, 25–27, 32, 34, 39,
44, 58, 59, 170, 189
electric motors, 58
electrification, 45
electrochemistry, 23
electromagnetic radiation, 24,
26–27, 29, 32
electromagnetic rotary devices, 23
electromagnetism, 23, 29, 190
electronics, 1, 34, 66, 73, 102, 114,
174, 183, 233, 234
electronic traded funds, 150
electrophoretic displays, 71
Elfers, William, 88, 100
employment, 115, 186, *191*,
191–193. *See also* jobs
energy, 58, 66, 94, 98, 126, 167
engineering, 195
"Entanglement in a Quantum
Annealing Processor" (D-Wave),
224
entertainment, 110
entrepreneurial dynamism, 9
entrepreneurs and entrepreneurism,
125; commercialization and, 167;
decline of, 10; deep science and, 36;
democratization of, 84; economic
dynamism and, 155; funding, *161*;
innovation and, 74; jobs and, 12,
194–195; productivity of, 73;
scientific breakthroughs and, 140;
venture capital and, 72–76, 74;
World War II and, 82
epigenetics, 180
Epogen, 139
EPROM tunnel oxide (ETOX) flash
memory cell, 244
equity: listings, 150, *151*; markets,
146; private equity industry, 75;

research, banks and, 146; risk premium, 131
error uncorrectability, 207
ETOX. *See* EPROM tunnel oxide flash memory cell
Evans, Oliver, 32
Eventbrite, 121
Ewald, Bo, 220
expected-utility theory, 131
Explorations in Quantum Computing (Williams), 212
explosives, 58
Exposition Universelle ("World's Fair"), 24
exposome, 180

Facebook, 16, 119, 121, 130, 162, 166, 185, 199, 200
Fairchild, Sherman, 86, 88
Fairchild Semiconductor, 88, 90
Faraday, Michael, 23–25, 27, 29, 31, 170
farming, 192
Farris, Haig, 212
fast-integer factorization, 206
FDA. *See* Food and Drug Administration
FDI. *See* foreign direct investment
federal budget, R & D and, 57
Federal Food, Drug, and Cosmetic Act, 68–69
federal regulatory agencies, 159
Fernald, John, 8
Feynman, Richard, 28, 211, 227
Feynmann, 206
Fidelity Ventures, 89, 95, 152
financial crisis (2008), 68, 158
financial markets, diversity breakdown in, 117, 147
Flanders, Ralph, 81, 83

Flexible Tubing, 88
Fong, Kevin, 201
Food and Drug Administration (FDA), 68
Ford, Horace, 83
foreign direct investment (FDI), 52
Fortune 500 companies, 141
Foster, Richard, 193–194
Foxconn, 172–173
"free money," 155
FreshDirect, 121
funding: crowdfunding, 163–165; entrepreneurial, 161; of R & D, 51–58, 53, 54; start-up businesses, 198; traditional, 161. *See also* venture capital
fusion energy, 58
"Future Postponed, The: Why Declining Investment in Basic Research Threatens a U.S. Innovation Deficit" (MIT) 57

gaming industry, 95
Gates, Bill, 166–167
Gave, Louis-Vincent, 172
GDP. *See* gross domestic product
Gell-Mann, Murray, 29
gene–environment interactions, 179–180
Genentech, 91, 96, 141
General Electric, 58, 59, 90
General Motors, 172
general-purpose technologies (GPTs), 18, 33–35, 35, 94, 190–191
genetic sequencing, 94
Genomics 2.0, 177–181
Gilder, George, 38, 155, 186
Global Settlement regulations, 146
global value chains (GVC), 52

Google, 162, 166, 173, 200,
218–219, 222, 224, 226
Gordon, Robert, 8–9, 30
government, 6, 51–58, 53, 54, 82
GPTs. *See* general-purpose
technologies
Grannis, Scott, 189
graphene, 176
gravitational collapse, 23
gravity, 27
Great Depression, 80–81
Great Inventions, 9
Great Recession, 162
green energy, 126
Greylock Capital, 88–89, 91, 100
Grinich, Victor, 90
Griswold, Merrill, 81
gross domestic product (GDP):
decline of, 12; growth of, 41,
48; living standards and, 40, 47;
quantum mechanics and, 45; R
& D and, 57, 58; venture capital
and, 4, 13
Guggenheim Partners, 227
GVC. *See* global value chains

hard science. *See* deep science
hardware, 15–16, 65, 199
Harris & Harris Group, 106, 107,
157
Harvard Business Review, 213
Hayden, Stone and Company, 90
heavenly bodies, 27
hedge funds, 128, 145, 147, 155
herding behavior, 117
"hero device," 64
Hewlett, William, 89–90
Hewlett-Packard Company (HP),
90, 241
high-density flexible circuits, 71
high-growth firms, 11

High Voltage Engineering
Corporation, 85, 86
Hilton, Jeremy, 216
Hoagland, Henry, 95
Hoerni, Jean, 90
"home runs," 96–97, 123, 169
Hopkins, Samuel, 32
Horowitt, Greg, 134, 201
HP. *See* Hewlett-Packard Company
Hu, Yaw Wen, 243
Hubble Space Telescope, 21, 242
human genome, 113, 172, 177
humanity, 30
Hwang, Victor, 134, 201
hydroelectric power plant, 25, 26

IBM. *See* International Business
Machines
immunotherapy, 179
incoherent tunneling, 210
income, 115, 186, 192, 201
incubation programs, 128
incumbents, 11–12
IND. *See* investigational new drug
Industrial Revolution, 12, 19, 21,
31, 40, 80; deep science and, 41,
170; economic dynamism of, 22;
economy and, 170; machines
and, 43, 189; manufacturing
and, 182; productivity and,
42–43
industry, institutionalization of, 75
information technology, 110
information theory, 19, 21, 29, 72,
183, 190
initial public offerings (IPOs), 141,
143, 160; Amgen Inc., 139;
deep science and, 185–186;
of Intel, 138; market, 142;
marketing of, 150; public
markets and, 144, 150;

small, 150, *151*, 152, 153;
venture capital and, 78, 142–144
inkjet technology, 241
innovation, 15, 16, 19, 30, 36; angel
funds, 104; business and, 60–61;
commercialization and, 108, 203;
continuum, 141; corporations
and, 58–63; deep science and,
17, 72, 109, 133, 137, 166–167,
174–176, 201, 202; disruptive,
73, 169; diversity and, 71–72;
diversity breakdown and, 162;
economic dynamism and, 165,
168, 191–192, 196; economy
and, 140, 204; ecosystem, 140,
159–160; employment and, 191;
entrepreneurs and, 74; hardware,
65; Jobs on, 105; legal aspect
of, 158–159; living standards
and, 50, 188; manufacturing
and, 72; marketplace and, 135;
modularity–maturity matrix.,
69, 70; nanotechnology and, 94;
outsourcing of, 195; process, 51;
as race, 140; R & D and, 61, 62,
106, 111; R & D payoff and, 50;
software, 102; Sputnik I satellite
and, 86; STEM and, 198; "strict
liability" and, 158; technology and,
38, 51, 62, 116, 169; uncertainty
and, 74; venture capital and, 79
innovation ecosystem, 106, 200;
deep science, 52, 147, *160*,
162–163, 203; R & D payoff
and, 161
In-Q-Tel, 222
Instacart, 121
institutional crowdfunding
platforms, 164
Institutional Venture Associates, 91
insurance companies, 84

Intel Corporation, 91, 138, *138*
intellectual property (IP), 31–33, 78,
216; IPR, 52, 212–213. *See also*
patents
intellectual property rights (IPR),
52, 212–213
intelligent machines, 171–174, 180,
183, 190
interference, 207, 208, 209
internal combustion engine, 34
International Business Machines
(IBM), 2
International SEMATECH, 238
Internet, 16, 19, 67, 72, 75, 102,
110–111, 118, 154, 168
inventions, 9, 19, 30, 46
Invesco Perpetual, 154
investigational new drug (IND),
68–69
investing, 122, 123
Investment Bankers Association
Conference, 81
investment banks, 146, 147,
148–149, 186
Investment Company Act, 84
investment process, 163
Ionics, 85–86
iOS, 126
IP. *See* intellectual property
iPhones, 128
IPOs. *See* initial public offerings
IPR. *See* intellectual property rights

Jet-Heet, 85
J. H. Whitney and Company, 83–84,
88
jobs, 192, 196–198; creation, 9–10,
12, 77, 191; entrepreneurs
and, 12, 194–195; loss of, 150;
market, 193; reallocation, *10*;
STEM-related, 196, *197*, 198

JOBS. *See* Jumpstart Our Business Startups Act of 2012

Jobs, Steve, 95, 105–106, 181, 202

Jumpstart Our Business Startups (JOBS) Act of 2012, 164–165

Kaplan, Sarah, 193–194

Kelvin, Lord, 29

Kennedy, John F., 129

Kepler, Johannes, 29

Kevlar, 59

"Killing the Golden Goose? The Decline of Science in Corporate R&D" (National Bureau of Economic Research), 60

Kim, Edward, 144, 145–146

Kiva Systems, 172

Kleiner, Eugene, 90

Kleiner, Perkins, Caufield and Byers, 90–91, 96

Kleiner Perkins, 90, 91, 94, 95, 96

Knowledge and the Wealth of Nations (Warsh), 38–39

Kurzweil, Raymond, 171, 172

Kwolek, Stephanie, 59

labor, 9; cost of, 39; machines and, 43, 172; market, 77, 192–193, 198; productivity, 43, 44; technology and, 39

Ladizinsky, Eric, 213, 215

Laplace, Pierre-Simon, 23, 29

Last, Jay, 90

law, administrative, 159

Lean Startup, The (Ries), 203

"LeMans," 95

Lerner, Steve, 241

life sciences, 103, 110, 178

lighting, 39, 45, 58

Lincoln Laboratory, 89

liquidity, microcapitalization and, 150

liquidity providers, 145

living standards, 36, 38; consumers and, 192–193; deep science and, 44–45; economic dynamism and, 47, 190, 191; GDP and, 40, 47; innovation and, 50, 188; lighting and, 45; machines and, 43, 44–45; technology and, 47 ; U.S. Constitution and, 188–189

Lockheed Martin, 219, 226, 241–242

long-tail distribution, 123

long-term stock exchange (LTSE), 203–204

Lorenz, Edward, 29

Los Alamos National Laboratory, 226

Lovelace, Ada, 21

LSI Logic, 238

LTSE. *See* long-term stock exchange

Lux Capital, 200–201

Lyft, 121

Ma, Jack, 167

Ma, Jeff, 124

M & A. *See* mergers and acquisitions

machines: age of, 36, 40; alternating-current, 43; dynamo, 24, 25, 34, 39, 58; electricity and, 26, 32, 44; Industrial Revolution and, 43, 189; intelligent, 171–174, 180, 183, 190; labor and, 43, 172; learning, 171, 178, 200, 202, 211, 218–219, 225, 228; living standards and, 43, 44–45; "Newtonian," 22, 23, 170; production and, 172; productivity and, 42–43, 44; "Quantum," 170

Macready, Bill, 216

macroeconomic trends, 135

Makers: The New Industrial Revolution (Anderson, C.), 181

Makimoto, Tsugio, 243
Malthus, Thomas Robert, 38
Manhattan Project, 83, 232
manufacturing, 64–65, 66, 69, 71,
    72, 135, 181–182, *182*, 238, 240
market capitalization, 13, 78
market inefficiencies, 117
market making, 145
marketplace, innovation and, 135
Markkula, A. C., Jr. ("Mike"), 95
Massachusetts Institute of
    Technology (MIT), 1, 81, 89
Massachusetts Investors Trust, 84
*Mathematical Principles of Natural
    Philosophy.* See *Philosophiæ
    Naturalis Principia Mathematica*
Mauboussin, Michael, 123, 155, 156
Maxwell, James Clerk, 23–25, 27,
    29, 31, 170
Mayfield Fund, 91, 95
media, 110
medical devices, 114–115
medicine, 94, 177–181, 183, 202–203
mega-funds, 103
MEMS. *See* microelectromechanical
    systems
mergers and acquisitions (M & A),
    160
metabolomics, 177–178, 180
Mezrich, Ben, 124–125
microbiome, 177–178
microcapitalization, 144–145, 150,
    152, 155, 163, 165–166, 187
microelectromechanical systems
    (MEMS), 241
microprocessors, 19, 39, 72, 138,
    168, 174, 190, 234
"micro VCs," 128
middle class, *103*, 189–190
missile development, 86
"Mission Innovation," 167

MIT. *See* Massachusetts Institute of
    Technology
modularity–maturity matrix, 69, *70*
molecular biology, 179
monetary system, 155
MoneyTree Report, 153
Moore, Gordon, 90, 91, 173–174,
    190, 234
Moore's law, 173–174, 190, 228, 233
Murray, Charles, 157
Musk, Elon, 65–66, 201
mutual funds, 128

Nadler, Marcus, 81
Naik, Ullas, 239
NAND flash memory circuit design,
    243, 245
nanomaterials, 71, 183
nanotechnology, 94, 98, 186–187,
    239, 241
Nantero, 114, 233; ASML and,
    237–238; CNTs, 176, 183–184,
    234–236, 238–240, 242, 245;
    commercialization at, 244,
    246–247; corporate partners at,
    246; founding of, 234–235, 248;
    Lockheed Martin and, 241–242;
    manufacturing of, 238, 240;
    NRAM technology, 176, 184,
    234–236, 239–240, 242–246, 248;
    patents of, 238, 246; R & D at,
    244; scalability at, 245; Series A
    financing round, 235–237; Series
    B financing round, 237–239; Series
    C financing round, 239–242; Series
    D financing round, 242–243;
    Series E financing round, 243–245;
    stock of, 236; SVTC Technologies
    and, 241; timeline, 247
NASA. *See* National Aeronautics
    and Space Administration

Nasdaq, 165
National Aeronautics and Space
 Administration (NASA), 86, 215,
 222, 226
National Market System (NMS),
 143, *151*
National Nanotechnology Initiative
 (NNI), 186–187
National Science Board, 198
National Science Foundation (NSF),
 196, 197
National Venture Capital Association
 (NVCA), 111, 144, 153
Nature, 221
nature, quantum mechanics and, 28
Nature Communications, 222
Neanderthals, 36
"Near-Term Quantum
 Computing Approach for Hard
 Computational Problems in
 Space Exploration, A" (D-Wave/
 NASA), 222
Neural Information Processing
 Systems, 219
neurobiology, 57
neuromorphic computing, 176
New England Council, 81, 83
Newport, Lisa, 145
Newton, Isaac, 19, 22, 23, 27, 29,
 31, 37
"Newtonian" machines, 22, 23, 170
Newton's laws, 37
Niagara Falls Power Company, 25,
 26
niobium, 221
NMS. *See* National Market System
NNI. *See* National Nanotechnology
 Initiative
nonphysical phenomena, 28
nonprofit organizations, R & D
 and, 55–56

nonreflective glass, 59
Nordhaus, William, 39
Noyce, Robert, 90, 91
NRAM technology, 176, 184, 234–
 236, 239–240, 242–246, 248
NSF. *See* National Science
 Foundation
NVCA. *See* National Venture
 Capital Association

Obama, Barack, 197–198
Office of Scientific Research and
 Development, U.S. (OSRD),
 82–83
Olsen, Kenneth, 2
1QBit, 225, 227, 228
online brokerage accounts, 142, 144
online brokerage commissions, 146
"open cloud," 127–128
open-source computing, 127
Origin of Wealth, The (Beinhocker),
 37
OSRD. *See* Office of Scientific
 Research and Development, U.S.
outsourcing, of innovation, 195

Packard, David ("Dave"), 89, 90
"passing the baton," 140
patents, 32–33, *33*, 60, 78, *182*; of
 D-Wave Systems, 216, 223, 227;
 of Nantero, 238, 246
Patient Protection, 158
Perkins, Tom, 90, 95
Pessimistic Scenario, 202
*Philosophiæ Naturalis Principia
 Mathematica* (*Mathematical
 Principles of Natural Philosophy*)
 (Newton), 19
Physical Review X, 224
physical sciences, 103, 110
physics, 19, 20, 27, 29, 174, 189

Pisano, Gary, 69, 70, 71, 91–92, 93, 94, 170, 175, 183
Planck, Max, 29
planetary motion, 29
"platform cloud," 127
poker, 123–125
policymakers, 7, 50, 73, 169, 194, 196, 198, 199, 203, 204
polymers, 59
"Pong," 95
positive spillovers, 4
"postwar social contract," 83
poverty, 190
prebiotics, 178
Precision Medicine, 177–181, 183, 202–203
PricewaterhouseCoopers, 153
Principia. See *Philosophiæ Naturalis Principia Mathematica*
printing, 43
private equity industry, 75
private sector, 198
probability theory, 23, 29
probiotics, 178
Producing Prosperity (Pisano & Shih), 69, 70, 71
product design, manufacturing and, 71
product development, from deep science, 18
production, 5, 172
productivity, 108, 115; definition of, 44; enhancement of, 199; of entrepreneurs, 73; growth of, 7, 8, 42; Industrial Revolution and, 42–43; inventions and, 9; labor, 43, 44; machines and, 42–43, 44; R & D and, 5–6, 50; slowdown of, 12, 14; technology and, 8
prospect theory, 131
prosperity, 16, 45, 199, 202

proteomics, 180
Prudent Man Rule, 78
public companies, 78, 152
public markets, 50, 141, 144, 150, 157, 165, 166
"Put Science to Work" program, 81

quadratic unconstrained binary optimization (QUBO), 216
QuAIL. *See* Quantum Artificial Intelligence Laboratory
quantum algorithm, 208, 209, 211
quantum annealer, 205, 221, 223, 225, 227, 229
"Quantum Annealing with Manufactured Spins" (D-Wave), 221
Quantum Artificial Intelligence Laboratory (QuAIL), 222, 226
quantum coherent phenomena, 205
quantum coherent tunneling, 210–211
quantum computing, 21, 47, 94, 175, 203, 205, 227; adiabatic model of, 214–215, 218; benefits of, 211; birth of, 206–207; bits in, 207–209; hardware, 213; memory in, 208, 209; power usage in, 208–209; solution time in, 211
quantum economy, 46
quantum effects, in D-Wave Systems, 207–211
quantum entanglement, 207, 209
quantum information technologies, 58
quantum machine learning, 218–219
"Quantum" machines, 170
quantum mechanics, 19, 27–28, 29, 31–33, 45–47, 173, 190
quantum phenomena, 208–209

quantum physics, 174
quantum processors, 207–208, 216
quantum revolution, 190
quantum science, 173, 205
quantum software ecosystem, 228
QUBO. *See* quadratic unconstrained
    binary optimization

Raam, Michael, 242
Radiation Laboratory, 89
radical venture capital, 93, 94
radio-frequency identification
    (RFID) tags, 241
Radio Research Laboratory, 90
railroads, 34, 58
*Rainforest, The: The Secret to
    Building the Next Silicon Valley*
    (Hwang & Horowitt), 134
Rainier chip, 220
Rao, Mohan, 237
R & D. *See* research and development
R & D payoff, 8–11, 13; companies
    and, 14; decreasing, 6, 7; deep
    science, 186–187; diversity
    and, 132; economic dynamism
    and, 105; innovation and, 50;
    innovation ecosystem and, 161;
    venture capital and, 6, 12, 188
reality, 27, 37
Regulation Fair Disclosure, 146
regulations, 157–159
relativity theory, 29, 37, 190
research and development (R & D):
    academia and, 53, 55, 56;
    computer science and, 14–15;
    corporations and, 52, 60; deep
    science and, 75, 111; diversity
    breakdown and, 187; D-Wave
    Systems, 230; economy and, 5,
    108; elasticity component in, 5;
    federal budget and, 57; funding

of, 51–58, 53; GDP and, 57, 58;
    government funding of, 51, 53,
    54; innovation and, 61, 62, 106,
    111; at Nantero, 244; nonprofit
    organizations and, 55–56;
    output and productivity of, 5–6;
    productivity and, 5–6, 50; of
    public companies, 78; R & D
    and, 6; ROI of, 61; spending and,
    79; spending by segment, 53, 54;
    trends in, 60, 61; venture capital
    and, 2, 4, 77; World War II and,
    132, 188. *See also* R & D payoff
resources, growth and, 37
return on investment (ROI), 61
Revenue Acts, 81
revenue growth, 136
RFID. *See* radio-frequency
    identification tags
Ries, Eric, 203–204
*Rise and Fall of American Growth,
    The* (Gordon), 9
risk, 3, 74, 82, 87
risk–return profile, 136
Roberts, Sheldon, 90
robotics, 58, 172, 200, 202, 211
Rock, Arthur, 88, 90, 91
Rockefeller Brothers Company,
    83–84, 89
ROI. *See* return on investment
Rond d'Alembert, Jean le, 22
Rose, Geordie, 212, 215, 220
Rothe, Tanya, 216
routine venture capital, 93
RSA public key cryptosystem,
    206–207
Rueckes, Thomas, 234

SandAire, 154
Sarbanes–Oxley Act of 2002, 142,
    157, 158

SBIC. *See* Small Business Investment Company program
Scandal of Money, The (Gilder), 155
Schmergel, Greg, 114, 234
Schmidt, Eric, 173
Schockley Semiconductor Laboratory, 90
Schumpeter, Joseph, 73, 169, 191
science, 7, 19, 30, 36–38, 53, 64, 195. *See also* deep science
science, technology, engineering, and math (STEM), 195–198, 197
Science and Engineering Indicators report, 198
Science: The Endless Frontier (Bush, V.), 82–83
scientific breakthroughs, entrepreneurs and, 140
scientific discovery, 30, 141
scientific progress, 1
SEC. *See* Securities and Exchange Commission, U.S.
SecondMarket, 165
Securities and Exchange Commission, U.S. (SEC), 84, 87, 163–165, 204
Segal, Brent, 234, 242
self-driving cars, 171–172, 199
Semiconductor Group, 237
semiconductors, 66, 71, 87, 88, 102, 114, 234, 238, 243. *See also* specific types
Sequoia Capital, 91, 94
Shang-Yi Chiang, 243
Shannon, Claude, 21, 29
Shih, Willy, 69, 70, 71
Shockley, William, 90
Shor, Peter, 206–207
Shor's algorithm, 206–207
short-termism, 60, 154–157
silicon, 234

Silicon Valley, 72, 75, 111–114, 120, 161, 184–185; deep science and, 129; employment in, 193; Moore's law and, 173–174; recalibration needed in, 204; social media and, 129
skill-luck continuum, 123
Small Business Innovation Research, 106
Small Business Investment Company (SBIC) program, 86, 88
small-capitalization stocks, 150
Smarter, Faster, Better (Duhigg), 123–124
smartphones, 128, 174
Snapchat, 8
Snyder Chemical Corporation, 85
social media, 8, 16, 119–120, 126–129, 163–165, 183, 187
society, 201
Sodhani, Arvind, 198–199
software, 67, 72, 119, 169, 185; Andreessen on, 109; applications, 16, 120; business model, 16–17; deep science and, 14–17, 200; diversity breakdown and, 118, 122; engineering, 15; industry, 127; innovation, 102; open-source computing, 127; quantum software ecosystem, 228; start-up business and, 199; technology and, 136; venture capital and, 13, 67, 109–110, 111, 126
solar system, 29
solid-state drive (SSD), 242–243
Sputnik I satellite, 86
SSD. *See* solid-state drive
staircase value appreciation, 133, 135, 137

Stanford University, 90
start-up business, 74, 77, 100, 119, 122, 164, 194–195, 198–199
steam engine, 36
STEM. *See* science, technology, engineering, and math
stock brokerage, 144
stock market, 116–118, 144–146, 152, 157
stock research, 146–147
Strategy + Business, 61–62
"strict liability," 158
success, measuring, 133–139
Success Equation, The (Mauboussin), 123
superconductivity, 227–228
superposition, 207–209
Sustainable Development Goals, UN, 194
Suster, Mark, 127, 128
Sutter Hill Ventures, 91
SVTC Technologies, 241
Swanson, Robert, 96
synthetic biology, 58
synthetic fibers, 59

tablets, 128
Takeuchi, Ken, 243
Tandem Computers, 90–91, 95, 125
taxes, 80–81, 158
technology, 22; biotechnology, 16, 67, 93–94, 102, 112–114, *113*, 115, 130, 168; chip, 15–16; commercialization of, 74, 115, 181; deep science and, 18–19, 30, 31, 33–34, 36–37, 39, 40, 48, 65, 67, 131, 140, 189, 192; diffusion rate of new, *156*, *157*; economic dynamism and, 30, 33, 47, 183; economy and, 35; employment and, 192; GPTs, 33–35; growth

of, *156*; information, 110; inkjet, 241; innovation and, 38, 51, 62, 116, 169; labor and, 39; living standards and, 47; nanotechnology, 94, 98, 186–187, 239, 241; NRAM, 176, 184, 234–236, 239–240, 242–246, 248; productivity and, 8; quantum mechanics and, 45, 46–47; science and, 7, 36, 38; software and, 136; transformative, 184; venture capital and, 1, 48, 86. *See also* machines
telecommunications, 72, 168
telegraph, 26, 34
telephone, 26, 34, 58, 157
telescopes, 21–22, 29
Teradyne, 87
Terman, Frederick, 89–90
Tesla, Nikola, 25, 28, 43, 129
Tesla electric car, 65–66
Tesla Motors, 201
Tesla polyphase system, 25
Tesla–Westinghouse power plants, 25
"Thermally Assisted Quantum Annealing of a 16-Qubit Problem," 223
thermodynamics, 29, 36
Thiel, Peter, 168
3D biology, 178
3D printing, 94, 177, 181–182
Thumbtack, 121
time-to-solution (TTS) metric, 229
time-to-target (TTT) metric, 229
Toshiba, 243
Tracerlab Incorporated, 85
traditional funding, *161*
"traitorous eight," 90
transformative technology, 184

transportation, 34
Treybig, James, 95
TTS. *See* time-to-solution metric
TTT. *See* time-to-target metric
Turnaround Scenario, 202–203
Twitter, 8, 16, 185
2D advanced nanomaterials, 176

Uber, 120–121, 122, 130, 152, 185
UN. *See* United Nations
uncertainty, innovation and, 74
unemployment, 196
unicorns, 13, 121, 130, 165, 166
United Nations (UN), 194
universe, science and, 19
university endowments, 84
UNIX servers, 127
USA Science and Engineering
  Festival, 195
USRA (Universities Space Research
  Association), 222, 226
utility patents, 32, 33

Valentine, Donald ("Don"), 91, 95
valuations, herding behavior and,
  117
value appreciation, staircase, *133*,
  135, *137*
value creation, *136*, *137*, 137–138,
  165, 166
venture capital: architectural, *93*,
  94; changing nature of, 98–102;
  companies and, 3, 3–4, 13,
  77–78; corporations and, 63, 66;
  deep science and, 106, 159–160,
  169–171, 175, 202; disruptive,
  93; diversity breakdown and,
  187, 200; economic dynamism
  and, 12–13, 14, 51, 75, 79,
  92, 116, 201; economy and,
  14, 72–79, 76; entrepreneurs

and, 72–76; environment, 116;
  evolution of, 80, 92; GDP and, 4,
  13; growth of, *101*; innovation
  and, 79; IPOs and, 78, 142–144;
  in media, 185; middle class, *103*;
  Pisano framework, *93*, 94, 170;
  policymakers and, 7; portfolios,
  97; postwar, 7; radical, *93*, 94; R
  & D and, 2, 4, 77; R & D payoff
  and, 6, 12, 188; reality *vs.*, *134*;
  representatives, 100; returns,
  97; risk and, 3, 74; routine, *93*;
  science and, 7; by sector, *112*,
  114; software and, 13, 67, 109–
  110, *111*, 126; technology and, 1,
  48, 86; trends, 75, 76, 109–115;
  types of, 91–96
venture funds, 99, 136–137
venture investing: definition of,
  93; diversity breakdown in,
  115–126, 129–130; economic
  dynamism and, 204; migration
  of, 110, 112, 200; model for,
  125; scenarios, 201–204; success
  in, 96–97
very-large-scale-integration (VLSI)
  chips, 237
"Vesuvius" chip, 220
video games, 168
VLSI Technology, 243
Voss, Jason, 150
VSLI. *See* very-large-scale-
  integration chips

Wall, Warren, 219
"Wanted: 600 Millions New Jobs"
  (Dell, M.), 194
Warsh, David, 38–39
Washington, George, 32
Watsons of International Business
  Machines (IBM), 88

Weild, David, 141–142, 144, 145–146
Westinghouse, George, 25
White, Mary Jo, 164
"Why Are IPOs in the ICU?" (Weild & Kim) 141–142
Wiens, Bob, 212
Williams, Colin P., 212, 215, 220
Wilson, Fred, 121–122
Woodford, 154
World Federation of Exchanges, 150

World's Fair. *See* Exposition Universelle
World War II, 58, 59, 82, 89, 132, 141, 188
Wozniak, Steve, 95, 181
writing skills, 30
Wythes, Paul, 91

Zagoskin, Alexandre, 212
Zuckerberg, Mark, 167